MW01601907

ANDROGENS

AND THE

AGING MALE

ANDROGENS
AND THE
AGING MALE

Proceedings of a workshop organized by the
International Health Foundation,
Geneva, December 1995

Edited by
B. J. Oddens and A. Vermeulen

The editors were assisted by Monique Boulet

The Parthenon Publishing Group
International Publishers in Medicine, Science & Technology

NEW YORK   LONDON

Published in North America by
The Parthenon Publishing Group Inc.
One Blue Hill Plaza, PO Box 1564, Pearl River
New York 10965, USA

Copyright © 1996 The Parthenon Publishing Group Ltd.

Published in the UK and Europe by
The Parthenon Publishing Group Ltd.
Casterton Hall, Carnforth, Lancs. LA6 2LA, UK

Library of Congress Cataloging-in-Publication Data
Androgens in the aging male: the proceedings of a workshop organized by the International Health
  Foundation, Geneva, Switzerland, December 1995/edited by B. J. Oddens and A. Vermeulen.
    p.   cm.
  Includes bibliographical references and index.
  ISBN 1-85070-763-4
  1. Androgens—Physiological effect—Congresses. 2. Androgens—Pathophysiology—Congresses.
3. Aging—Endocrine aspects—Congresses. 4. Aged men—Health and hygiene—Congresses.
I. Oddens, B. J. (Bjorn J.) II. Vermeulen, A. (Alex), 1927-. III. International Health Foundation.
  [DNLM: 1. Androgens—congresses. 2. Aging—congresses. WJ 875 A574 1996]
QP572. A5A535   1996
612.6′1—dc20
DNLM/DLC
for Library of Congress   96-29358
                                                                                                              CIP

British Library Cataloguing in Publication Data
Androgens in the aging male: the proceedings of a workshop organized by the International Health
  Foundation, Geneva, Switzerland, December 1995
  1. Androgens   2. Generative organs, Male   3. Aged men—Health and hygiene
I. Oddens, B. J.   II. Vermeulen, A.
612.6′1′0846
ISBN 1-85070-763-4

Typeset by AMA Graphics Ltd., Preston, Lancashire
Printed and bound by Bookcraft (Bath) Ltd., Midsomer Norton, UK

Contents

List of participants

J. Bain MD
Associate Professor
Endocrinology-Fertility and
    Reproduction
Mount Sinai Hospital
600 University Avenue, Suite 781
Toronto, Ontario M5G 1X5,
Canada

Dr W. E. Bergink
Programme Manager Reproductive
    Medicine
N.V. Organon
KA2012
P.O. Box 20
5340 BH Oss, The Netherlands

Mrs M. J. Boulet
Scientific Editor
International Health Foundation
Avenue de Broqueville 116, B9
1200 Brussels, Belgium

Priv.-Doz. Dr K. Christiansen
Institut für Humanbiologie
Universität Hamburg
Allende-Platz 2
20146 Hamburg, Germany

Dr J. Ginsburg
Consultant Endocrinologist
Royal Free Hospital School of
    Medicine
Pond Street
London NW3 2QG, UK

Professor Dr L. J. G. Gooren
Department of Andrology
AZVU
P.O. Box 7057
1007 MB Amsterdam,
The Netherlands

Professor S. M. Haffner
Department of Medicine
Division of Clinical Epidemiology
The University of Texas
Health Science Center at San Antonio
7703 Floyd Curl Drive
San Antonio, Texas 78284-7873, USA

Professor Dr J. M. Kaufman
Department of Endocrinology
University Hospital
De Pintelaan 185
9000 Gent, Belgium

Professor C. Longcope
Department of OB/GYN
University of Massachusetts Medical
    School
55 Lake Avenue, North,
Worcester, MA 01655, USA

Professor Dr G. Lunglmayr
Abteilung für Urologie
Krankenhaus Mistelbach
Liechtensteinstrasse 67
A-2130 Mistelbach/Zaya, Austria

Professor A. Morales, FRCS(C)
Department of Urology
Queen's University
Kingston General Hospital
76 Stuart Street
Kingston, Ontario K7L 2V7, Canada

B. J. Oddens, MD PhD
Research Director
International Health Foundation
Avenue de Broqueville 116, B9
1200 Brussels, Belgium

E. S. Orwoll, MD
Chief, Endocrinology & Metabolism
Bone and Mineral Research Unit
Portland VA Medical Center (111P)
and Oregon Health Sciences University
P.O. Box 1034
Portland, OR 79225, USA

Mrs H. Rilly
17 Amberlane
Esher
Surrey KT10 8DZ
London, UK

Professor R. C. Schiavi
The Mount Sinai Medical School
Department of Psychiatry
1 Gustave L. Levy Place (Box 1084)
New York, NY 10029-6574, USA

Professor Dr F. H. Schröder
Department of Urology
Erasmus University
P.O. Box 1738
3000 DR Rotterdam, The Netherlands

Dr D. Simon
INSERM (Institut National de la Santé
et de la Recherche Médicale)
Unité 21 – Unité de Recherches
Cliniques et Epidémiologiques
16, Avenue Paul-Vaillant-Couturier
94807 Villejuif, France

J. L. Tenover, MD PhD
Associate Professor of Medicine
Division of Gerontology and Geriatric
Medicine
Emory University School of Medicine
Wesley Woods Health Center, 2nd
Floor
1841 Clifton Road, N.E.
Atlanta, Georgia 30329-5102, USA

Professor Dr J. H. H. Thijssen
Department of Endocrinology
Academisch Ziekenhuis Utrecht
P.O. Box 85500
3508 GA Utrecht, The Netherlands

Professor Dr A. Vermeulen
Maaltemeers 33
9052 Zwijnaarde-Gent, Belgium

Emeritas Professor Dr D. de Wied
Utrecht University
Rufolf Magnus Institute for
Neurosciences
Universiteitsweg 100
3584 CG Utrecht, The Netherlands

Preface

D. de Wied

It is generally acknowledged that men, unlike women, do not experience an abrupt decline in gonadal function as early as in the fifth decade. Rather, a gradual decline in androgen levels has consistently been observed with advancement of age, independently of previous diseases and resulting in some cases in considerably lowered levels at older age. The terms 'male menopause' and 'andropause' suggest an abrupt phenomenon and an onset of complaints and health problems related to a sudden deprivation of sex hormones, and may therefore cause confusion. It seems preferable to speak of declining androgens with advancement of age and to focus on gradually progressive problems which might be possibly related to this decline. It has been hypothesized that in some men a certain relative androgen deficiency might develop with aging and this hypothesis warrants further consideration.

The potential consequences of the decline of bioactive androgen levels with age were the central topic of a workshop organized by the International Health Foundation in Geneva, Switzerland, in December 1995. Various experts were invited to present and discuss their research and views on the issue. The presentations and discussions are summarized in these proceedings. Research has suggested that androgens play distinct roles in aspects of sexual behavior, bone metabolism, body composition such as muscle and fat mass and fat distribution, cognitive functioning, mood, well-being, the development (or prevention?) of cardiovascular disease and prostate hyperplasia. The extent to which an age-dependent decline in androgens leads to health problems which might affect or impair the quality of life, is under debate and is actually further discussed in the various chapters of this book. Some studies have evaluated the effects of androgen supplementation and they are extensively discussed as well. Accordingly, this monograph aims at presenting the state of the art with respect to androgen functioning and aging in men.

The workshop of which the proceedings are covered here was not the first organized by the International Health Foundation. The Foundation is a private research organization, acting under the supervision of a Board of Trustees. It focuses in its research and information provision on reproductive health issues, such as contraception, menopause and infertility.

Since its inception in 1969, 14 workshops have been organized, most of them dealing with various aspects of family planning and menopause. Generally, the topics of the workshops relate to issues adjacent to the daily activities of the Foundation, in particular where there exist controversies, unclarity or lack of insights. Although the Foundation has focused so far more often on issues related to women's health, the Board and staff considered it worthwhile devoting the anniversary workshop in the 25th year of the Foundation's inception to androgen functioning in the aging male.

The contributors were asked to extend their views to what might possibly be the future of research and health care, so that hints as to immediate research priorities may be clearly discernible. We hope that these proceedings will be of interest and help for those who are dedicated to health and quality of life of aging men, and may contribute to a wide appraisal of the intriguing research that is currently being carried out in this connection.

David de Wied
Chairman, Board of Trustees
International Health Foundation

1

Declining androgens with age: an overview

A. Vermeulen

Whereas the persistence of fertility in elderly males might suggest better vitality of men in comparison to women, the 8-years shorter life expectancy of men proves that this is not the case and should constitute a strong rationale for studying the aging phenomena in men. Men indeed represent the weaker sex!

While the role of hypo-estrogenism in the pathogenesis of the menopausal and aging phenomena in women is well established, the role of androgens in the aging process of the male is still controversial. Nevertheless, carefully collected data concerning the influence of age on androgen plasma levels, obtained not only in cross-sectional (for review see ref. 1) but also in longitudinal studies[2] involving large groups of subjects, are now available and they have shown, beyond any discussion, that in healthy men and independently of previous diseases an important age-associated decrease in the levels of bioactive androgens does occur (Table 1). Nevertheless it is evident that an additional decrease may occur as a sequel of several disease states[3]. Moreover, many physiological and pathological, as well as environmental, factors influencing the androgen plasma levels have been identified[4].

Also the mechanisms of the age-associated decrease of the endocrine testicular function have been largely elucidated and we now know that this decrease has essentially a primary testicular origin but that important changes also occur at the hypothalamo–pituitary level[5].

Although our knowledge concerning the relationship between androgens and the aging process in males has increased tremendously in recent years, many problems, nevertheless, remain unsolved. For example, why does sex hormone binding globulin (SHBG) binding capacity increase in elderly men, although insulin levels, which are known to decrease SHBG

Table 1 Mean plasma sex hormone levels in healthy men, by age (SD in parentheses); $n = 249$

Age (years)	n	T	FT	SHBG	DHEA	DHEAS	A	E_2
25–34	45	21.38	0.428	35.5	15.91	6.44	3.85	136.8
		(5.90)	(0.098)	(8.8)	(6.05)	(2.29)	(1.25)	(50.4)
35–44	22	23.14	0.356	40.1	12.65	6.02	3.81	134.2
		(7.36)	(0.043)	(7.9)	(3.69)	(2.18)	(1.01)	(56.3)
45–54	23	21.02	0.314	44.6	11.31	4.75	3.36	142.3
		(7.37)	(0.075)	(8.2)	(5.39)	(2.62)	(0.90)	(37.1)
55–64	43	19.49	0.288	45.5	10.20	3.25	4.66	128.7
		(6.75)	(0.073)	(8.8)	(5.21)	(1.48)	(1.28)	(40.4)
65–74	47	18.15	0.239	48.7	7.71	2.65	4.47	132.3
		(6.83)	(0.078)	(14.2)	(4.15)	(1.68)	(2.17)	(38.2)
75–84	48	16.32	0.207	51.0	5.39	1.15	2.18	138.7
		(5.85)	(0.081)	(22.7)	(2.76)	(0.52)	(1.49)	(43.4)
85–100	21	13.05	0.186	65.9	3.18	1.23	1.85	136.4
		(4.63)	(0.080)	(22.8)	(0.69)	(0.52)	(0.91)	(39.5)

All values in nmol/l, except for DHEAS in μmol/l and E_2 in pmol/l. T, testosterone; FT, free testosterone; SHBG, sex hormone binding globulin; DHEA, dehydroepiandrosterone; DHEAS, dehydroepiandrosterone sulfate; A, androstenedione; E_2, estradiol

Unpublished findings from Vermeulen and Kaufman

synthesis and plasma levels, increase with aging? What are the clinical consequences of the decrease in testosterone levels and is there any correlation between androgen levels and symptoms such as tiredness, lack of energy, decrease in muscle strength, sleep disorders, decrease in libido and sexual activity or impotence, which to various degrees, accompany aging in men? Is there a causal relationship between decreasing androgen levels and senile osteoporosis? What is the exact role of testosterone in the pathogenesis of prostatic hyperplasia? While prostatic carcinoma is known to be androgen dependent, why is it that at equal frequency of latent (infraclinical) prostatic carcinoma and rather similar testosterone levels in Japanese and Western men, frequency of clinical carcinoma of the prostate in Japanese men is only one-tenth of the frequency in Western Europe or the USA? Is there any correlation between the relative immunodeficiency of the elderly and the decreasing androgen levels? What is the role of the androgens in the pathogenesis of atherosclerosis? Is the higher frequency of cardiovascular disease in men compared to premenopausal women related to the higher androgen levels in men? Is the estradiol to testosterone ratio a parameter of cardiovascular risk? How safe is androgen hormone replacement therapy in elderly men? These are a few of the many questions that remain unanswered and it is hoped that the workshop, the contributions of which have been covered in these proceedings, will give an answer to at least some of these questions.

Recently, a much neglected androgen steroid has attracted renewed interest in connection with aging, namely dehydroepiandrosterone (DHEA) and its sulfate (DHEAS). As no other contributor deals specifically with the possible role of this steroid in the aging process, I will expand somewhat on this problem.

DHEA and DHEAS, secreted almost exclusively by the adrenal cortex, are quantitatively the most important steroid hormones of the human organism, being secreted at a rate of about 35 mg/day in young adults. They are interconvertible in plasma: within 30 min of intravenous (i.v.) injection, 90% of DHEA is conjugated, whereas about 70% of DHEAS is converted to DHEA[6]. DHEAS has a half-life of approximately 11 h and a metabolic clearance rate of about 15 l/24 h. DHEA, on the other hand, is secreted almost synchronously with cortisol; it has a short half-life (about 25 min) and hence a high metabolic clearance rate of 2000 l/24 h. However, whereas cortisol plasma levels and secretion rates (per m²) remain constant throughout the entire lifespan, DHEA and DHEAS levels and production rates decrease by about 2%/year, such that at age 80 years plasma levels are only 20% of levels at age 20 years[7]. Moreover, while the nyctohemeral variations in cortisol levels remain uninfluenced by the aging process, the nyctohemeral variation of DHEA levels, which have an amplitude of 40% in young adults, have almost completely disappeared in men over 70 years old[7]. While the increase of DHEA levels after stimulation with adrenocorticotropic hormone (ACTH) indicates that the latter is an important regulator of DHEA levels, the age-associated decrease of DHEA(S) levels in the presence of unchanged cortisol levels suggests that other regulators play an important role. Although Parker and Odell[9] isolated a pituitary polypeptide which they report to specifically stimulate DHEA(S) secretion, other authors have not confirmed these findings and the specific adrenal androgen-stimulating hormone still remains elusive. Also, the role of insulin, which according to several authors[10,11] would decrease DHEA(S) plasma levels, is still controversial. Not only the factors regulating DHEA(S) secretion, but also the physiological role of DHEA(S) remains largely unknown. According to Morales and co-workers[12] DHEA may be viewed as a multifunctional steroid with protective roles in many aspects of cellular well-being, especially age-associated defects.

Experiments performed as long as 50 years ago have shown that DHEA(S) is a weak androgen, which in peripheral tissues is partially transformed in androstenedione and testosterone as well as in estrone[13]. 17β reduction of DHEA yields androstenediol, which binds to both the androgen and estrogen receptor. The impressive age-associated decrease of DHEA(S) levels leads to the speculation that their plasma level might be an excellent parameter of physiological aging, and that in several disease states, DHEA(S) levels are decreased in comparison to levels in healthy men of similar age[14]. As in elderly men, ACTH stimulation results in a normal increase of 17-hydroxyprogesterone (17OHP) levels whereas DHEA levels remain largely unchanged[15], Carlström[16] suggested the use of $\Delta17OHP/\Delta DHEA$ ratio as a parameter of physiological aging.

Data of Barrett-Connor and colleagues[17] suggested that men with low DHEAS levels were much more at risk of cardiac infarction and death from any cause within the next 12 years, than men with higher DHEAS levels; these data were however not confirmed later[18]. Nevertheless, data of Herrington[19] suggest that, in men, low plasma levels of DHEAS may facilitate, and high levels retard, the development of coronary atherosclerosis. Hautanen and co-workers[20], on the contrary, reported that men with coronary artery disease had higher DHEAS levels than controls. Similarly, Barrett-Connor and Khaw[21] observed high DHEAS levels in women to be associated with increased frequency of cardiovascular disease[19].

Helzlsouer and colleagues[22] reported that low DHEAS levels are correlated with an increased frequency of breast carcinoma. Gordon and co-workers[23] on the other hand reported that DHEA, but not DHEAS, levels were significantly higher in postmenopausal women who, later (> 9 years) developed breast carcinoma. The significance of this observation, is however questionable, as DHEA levels show important nyctohemeral variations. Ebeling and Koivisto[24] suggest that in premenopausal women, low DHEA levels would increase the risk of breast carcinoma, whereas in postmenopausal women high DHEA levels would increase the risk. They suggest that in premenopausal women, low DHEA would leave the estrogen effect on tumor cell growth unopposed, whereas in postmenopausal women high DHEA levels would act as an estrogen, possibly via transformation to androstene, 3β,17β-diol. Reed and Ghilchik[25] claim that low DHEA levels stimulate the progression of T helper cells to the Th2 phenotype cells, which secrete interleukin-6 (IL-6) that has been shown to stimulate estrogen synthesis in cancer cells. It is difficult to draw a final conclusion from these data with regard to the effect of DHEA(S) on breast carcinoma.

Thomas and Weigle[26] observed a correlation between low DHEA levels and the decline of immune competence during aging. It has also been reported that DHEAS levels were inversely related to the presence of organic brain syndrome but not to the 1-year mortality rate[27]. As to DHEAS levels in Alzheimer's disease, data are controversial[28–31]. It should be recalled that DHEA(S) are neurosteroids acting directly on neural membranes; they are synthesized in the brain independently of classical steroidogenic organs[32] and act as γ-aminobutyric acid (GABA) antagonists in the central nervous system, resulting in analeptic and memory-enhancing effects. Moreover, they promote neuronal growth and regeneration[33].

The age-associated decrease of DHEA(S) levels leads to the speculation that normalization of these levels by exogenous substitution might have favorable effects on general well-being and act as a kind of juvenile hormone. Exogenous DHEA, at a dose of 50 mg/day, which yields physiological levels as observed in young healthy persons, was reported to improve general well-being, a 20% increase in insulin-like growth factor-I (IGF-I) levels being the major change in the biological parameters[12]. However, in women, this dose resulted in a moderate increase in testosterone levels and it is not clear whether the subjective impression of improvement of general well-being is not the consequence of the increase in the sex hormone levels. Higher doses of DHEA (300–1600 mg/day) induced a decrease of plasma cholesterol and a 30% decrease in body fat without weight change, indicating an increase in muscle mass; in women these doses induce however, an important increase in testosterone as well as in estradiol levels[34–36]. Araneo *et al.*[37] reported that administration of DHEA prior to immunization of elderly volunteers with influenza vaccine resulted in a significant increase in the percentage of women showing a fourfold increase in the antibody titer. DHEAS administration would also increase insulin resistance and modulate immune function in postmenopausal women stimulating natural killer lymphocyte activity[38].

Studies in animals revealed some additional effects of DHEA(S). Administration of DHEA to mice or rats inhibits experimentally induced tumors of breast, colon, skin, or liver[39]; in rabbits, DHEA reduces the development of vascular stenosis in heterotopic heart transplants[40]. Moreover, in aged mice, supplementation with DHEAS restores the normal responsiveness to growth factors and the capacity to elicit normal immune responses to foreign antigens[41], whereas in spontaneously obese dogs[42] as well as in obese rats[43], DHEA administration causes a reduction of body weight by reducing food intake.

In vitro studies have shown that DHEAS at physiological concentration reduces oxygen uptake by monocytes, suggesting that it may down-regulate the earliest stages of the inflammatory immune response. Lardy[44] observed that DHEA(S) increases the amounts of thermogenic enzymes in liver mitochondria several-fold. Finally, in *in vitro* experiments, DHEA inhibited arachidonic acid-induced thromboxane B2 synthesis and platelet aggregation[45].

From all these studies, it appears that DHEA(S), under different experimental conditions, exhibits a large array of biological effects which, if similar effects were observed in men, would possibly warrant DHEA(S) replacement therapy in elderly men and women.

A word of caution, however. Although many observations, both in animals and man, support the idea that DHEA(S) may have a biological role in men, it should be recalled that DHEA(S) production in mice and rats is minimal or absent. Extrapolation of results obtained in these animals to men is therefore at least hazardous, whereas the effects obtained *in vitro* indicate possibilities, but by no means prove the existence of similar effects under *in vivo* conditions. Finally, it should be recalled that aging is a multifactorial process and that the decrease in sexual hormones is only one aspect of the endocrine changes of the aging process, which is also accompanied by an important decrease in growth hormone and IGF-I secretion as well as with increased insulin levels, a consequence of the relative insulin resistance characterizing the aging process.

REFERENCES

1. Vermeulen, A. (1991). Androgens in the aging male – clinical review 24. *J. Clin. Endocrinol. Metab.*, **73**, 221–4
2. Pearson, J. D., Blackman, M. R., Metter E. J., Wachawiw, Z., Carter, H. B. and Harman, S. M. (1995). Effect of age and cigarette smoking on longitudinal changes in androgens and SHBG in healthy men. Abstracts of the 77th Annual Meeting of the Endocrinological Society, Abstract P2–129

3. Semple, G., Gray, C. E. and Beastall, G. H. (1987). Male hypogonadism, a non specific consequence of illness. *Q. J. Med.*, **243**, 601–7

4. Vermeulen, A. (1993). Environment, human reproduction, menopause and andropause. *Environm. Health Persp.*, Suppl., **101**, 91–100

5. Vermeulen, A. and Kaufman, J. M. (1992). Editorial: role of the hypothalamo–pituitary function in the hypoandrogenism of healthy aging. *J. Clin. Endocrinol. Metab.*, **74**, 1226A–C

6. Bird, C., Master, V. and Clark, A. (1984). DHAS: Kinetics and metabolism in normal young men and women. *Clin. Invest. Med.*, **7**, 119–24

7. Vermeulen, A. (1996). DHEA(S) and aging. *Ann. N.Y. Acad. Sci.*, **774**, 121–7

8. Deslypere, J. P. and Vermeulen, A. (1984). Leydig cell function in normal men: effects of age, life style, residence, diet and activity. *J. Clin. Endocrinol. Metab.*, **59**, 955–62

9. Parker, L. N. and Odell, W. D. (1980). Control of adrenal androgen secretion. *Endocr. Rev.*, **1**, 392–410

10. Nestler, J. E., Usiskin, K. S., Barlascini, C. O., Welty, D. F., Clore, J. N. and Blackard, W. G. (1989). Suppression of serum dehydroepiandrosteronesulfate levels by insulin: an evaluation of possible mechanisms. *J. Clin. Endocrinol. Metab.*, **69**, 1040–6

11. Hubert, G. D., Schriöck, E. D., Givens, J. R. and Buster, J. E. (1991). Suppression of circulating $\Delta 4$ androstenedione and dehydroepiandrosterone-sulfate during oral glucose tolerance test in normal females. *J. Clin. Endocrinol. Metab.*, **73**, 781–4

12. Morales, A. J., Nolan, J. J., Nelson, J. C. and Yen, S. S. C. (1994). Effects of replacement dose of dehydroepiandrosterone in men and women of advancing age. *J. Clin. Endocrinol. Metab.*, **78**, 1360–7

13. Horton, R. and Tait, J. (1967). The *in vivo* conversion of dehydroisoandrosterone to plasma androstenedione and testosterone in men. *J. Clin. Endocrinol. Metab.*, **27**, 79–88

14. Hedman, M., Nilsson, E. and de la Torre, B. (1992). Low blood and synovial fluid levels of sulphoconjugated steroids in rheumatoid arthritis. *Clin. Exp. Rheumatol.*, **10**, 25–30

15. Vermeulen, A., Deslypere, J. P., Schelfhout, N., Verdonck, L. and Rubens, R. (1992). Adrenocortical function in old age. Response to acute ACTH stimulation. *J. Clin. Endocrinol. Metab.*, **54**, 187–92

16. Carlström, K. (1994). Andropause, fact or fiction? In Berg, G. and Hammar, H. (eds.) *The Modern Management of the Menopause. A Perspective for the 21st*

Century, Proceedings VIIth International Congress on the Menopause, pp. 567–77. Stockholm, 1993. New York, London: Parthenon Publishing Group

17. Barrett-Connor, E., Khaw, K. T. and Yen, S. S. C. (1986). A prospective study of dehydroepiandrosteronesulfate, mortality and cardiovascular disease. *N. Engl. J. Med.*, **315**, 1519–24

18. Birkenhäger-Gillesse, E. C., Derksen, J. and Lagaay, A. M. (1994). Dehydroepiandrosteronesulfate (DHEAS) in the oldest old, age 85 and over. *Ann. N.Y. Acad. Sci.*, **719**, 543–52

19. Herrington, D. M. (1996). DHEA and coronary atherosclerosis. *Ann. N.Y. Acad. Sci.*, **774**, 271–80

20. Hautanen, A., Mänttärri, M., Manninen, V., Tenkanen, L., Huttunen, J. K., Frick, M. H. and Adlercreutz, H. (1994). Adrenal androgens and testosterone as coronary risk factors in the Helsinki heart study. *Atherosclerosis*, **105**, 191–200

21. Barrett-Connor, E. and Khaw, K. T. (1987). Absence of an inverse relation of dehydroepiandrosterone sulfate with cardiovascular mortality in postmenopausal women. *N. Engl. J. Med.*, **317**, 711 (letter)

22. Helzlsouer, K. J., Gordon, C. B., Alberg, A., Bush, T. L. and Comstock, G. W. (1992). Relationship of prediagnostic serum levels of DHEA and DS to the risk of developing premenopausal breast cancer. *Cancer Res.*, **52**, 1–4

23. Gordon, G. B., Bush, T. L., Helzlsouer, K. J., Miller, S. R. and Comstock, G. W. (1990). Relationship of serum levels of dehydroepiandrosterone and dehydroepiandrosterone sulfate to the risk of developing postmenopausal breast cancer. *Cancer Res.*, **50**, 3859–62

24. Ebeling, P. and Koivisto, V. A. (1994). Physiological importance of dehydroepiandrosterone. *Lancet*, 343, 1479–81

25. Reed, M. J. and Ghilchik, M. W. (1995). Risk of breast cancer after renal or cardiac transplantation. *Lancet* (letter), **346**, 1422–3

26. Thomas, M. L. and Weigle, W. O. (1989). The cellular and subcellular basis of immunosenescence. *Adv. Immunol.*, **46**, 221–61

27. Rudman, D., Shetty, K. R. and Mattson, V. (1990). Plasma dehydroepiandrosteronesulfate in nursing home men. *J. Am. Geriatr. Soc.*, **38**, 421–7

28. Sunderland, N., Merill, C. R., Harrington, M. G., Lawlor, B. A., Molchan, S. B., Martinez, R. and Murphy, D. L. (1989). Reduced plasma dehydroepiandrosterone concentration in Alzheimer's disease. *Lancet*, **ii**, 570

29. Leblhuber, F., Windhager, E., Reisecker, F., Steinparz, F. X. and Dienstl, E. (1990). Dehydroepiandrosteronesulphate in Alzheimer's disease. *Lancet* (letter), **336**, 449

30. Cuckle, H., Stone, R., Smith, D., Wald, N., Brammer, I., Hajimohamme-dreza, H., Levy, R., Chard, T. and Perry, L. (1990). Dehydroepiandrosterone sulphate in Alzheimer's Disease. *Lancet* (letter), **336**, 449–50

31. Schneider, L. S., Hinsey, M. and Lyness, S. (1992). Plasma dehydroepiandrosterone sulfate in Alzheimer's disease. *Biol. Psychiatr.*, **34**, 867–70

32. Robel, P. and Baulieu, E. E. (1994). Neurosteroids: biosynthesis and function. *Trends Endocrinol. Metab.*, **5**, 1–8

33. Majewska, M. D. (1996). Neuronal activities of DHEAS. Possible holes in brain development, aging, memory, and effect. *Ann. N. Y. Acad. Sci.*, **774**, 111–20

34. Mortola, J. F. and Yen, S. S. C. (1990). The effects of oral dehydroepiandrosterone on endocrine-metabolic parameters in postmenopausal women. *J. Clin. Endocrinol. Metab.*, **71**, 696–704

35. Buster, J. E., Casson, P. R., Straugh, A. B., Debra Dale R. N., Umstot, M. S., Chiamori, N. and Abraham, G. (1992). Postmenopausal steroid replacement with micronized dehydroepiandrosterone: preliminary oral bioavailability and dose proportionality studies. *Am. J. Obstet. Gynecol.*, **166**, 1163–70

36. Nestler, J., Barlascini, C. O., Clore, J. N. and Blackard, W. G. (1988). Dehydroepiandrosterone reduces serum low density lipoprotein levels and body fat but does not alter insulin sensitivity in normal men. *J. Clin. Endocrinol. Metabol.*, **66**, 57–61

37. Araneo, B., Dowell, T., Woods, M. L., Daynes, R., Judot, M. and Evans, T. (1996). DHEAS as an effective vaccine adjuvant in elderly humans: proof of principle studies. *Ann. N.Y. Acad. Sci.*, **774**, 232–48

38. Casson, P. R., Andersen, R. N., Herrod, H. G., Stentz, F. B., Stroughn, A. B., Abraham, G. E. and Buster, J. E. (1993). Oral dehydroepiandrosterone in physiologic doses modulates immune function in postmenopausal women. *Am. J. Obstet. Gynecol.*, **169**, 1536–9

39. Schwarz, A. G. and Pashko, L. L. (1996). Mechanism of cancer preventive action of DHEA: role of glucose-6-phosphate dehydrogenase. *Ann. N.Y. Acad. Sci.*, **774**, 180–86

40. Eich, D. M., Nestler, J. E., Johnson, D. E., Dworkin, G. H., Daijin, K. O., Wechsler, A. S. and Hess, M. L. (1993). Inhibition of accelerated atherosclerosis with dehydroepiandrosterone in the heterotopic rabbit model of cardiac transplantation. *Circulation*, **87**, 261–9

41. Daynes, R. A., Dudley, D. S. and Araneo, B. A. (1990). Regulation of murine lymphokine production *in vivo*. II Synthesis by helper T cells. *Eur. J. Immunol.*, **20**, 793–802

42. Kurzman, E., Macewen, G. and Haffa, L. M. (1990). Reduction in body weight and cholesterol in spontaneous obese dogs by dehydroepiandrosterone. *Int. J. Obesity*, **14**, 95–103

43. Mohan, P. F., Ihnen, J. S., Levin, B. E. and Cleary, M. P. (1990). Effects of dehydroepiandrosterone treatment in rats with diet inducing obesity. *J. Nutr.*, **120**, 1103–14

44. Lardy, H. A. (1996). Induction of thermogenic enzymes by DHEA and its metabolites. *Ann. N.Y. Acad. Sci.*, in press

45. Jesse, R. L., Loesser, K., Eich, D. M., Quian, Y. Z., Hess, M. L. and Nestler, J. E. (1996). Dehydroepiandrosterone inhibits human platelet aggregation *in vitro* and *in vivo*. *Ann. N.Y. Acad. Sci.*, **774**, 281–93

DISCUSSION

Gooren: Do you think there is a relationship between insulin levels and the increase of sex hormone-binding globulin (SHBG) levels? Sometimes you read that high portal insulin levels contribute to an increase of SHBG production in the liver.

Vermeulen: Certainly, insulin levels influence SHBG levels in the sense that high insulin levels decrease SHBG. In this way, we try to explain the decrease in testosterone levels in obese men: these men have increased insulin resistance and low SHBG levels, low testosterone levels and, as long as the obesity is not too important, free testosterone levels in the normal range. Only with extreme obesity does free testosterone also decrease. Furthermore, in poly-cystic ovarian syndrome, there is insulin resistance and generally a decrease in SHBG levels. In elderly men, however, we see a certain degree of insulin resistance associated with higher insulin levels, but nevertheless, SHBG levels are rather higher than normal. Therefore, the increase in SHBG levels observed in elderly men cannot be explained by the insulin levels. The decreases in growth hormone or IGF–I levels probably play a role in the increase of SHBG levels.

Thijssen: You mentioned, indeed correctly, that DHEA is, in terms of evolution, a very late hormone. Only higher primates and humans have

DHEA, implying that since there is hardly any DHEA in rats, we lack an animal model for DHEA. Does the fact that DHEA appeared quite late in evolution give you any clues to speculate on its functions?

Vermeulen: Well, I have presented data suggesting a possible protective effect against atherosclerosis, and of a role in the immune processes in elderly men. However, in my opinion, most of the studies published so far are really preliminary, and we cannot draw any final conclusions about the possible effects of DHEA in men. In this respect, I must admit that I as somewhat surprised that in some countries, among which the United States is one, DHEA is already very popular as a supplemental therapy to prevent aging phenomena[1]. I think it is much too early to advocate such practice on the basis of the current data. I do not have any explanation as to the mechanism or reasons that DHEA is present only in higher animals.

Longcope: I have two comments. You focused on DHEA and its sulfate in relation to aging. However, their plasma levels are terribly low from birth onwards to practically 8 years of age, in fact even lower than observed with aging. Nevertheless, those individuals are not prone to the diseases of aging. So, there must obviously be many multifactorial variables at work here. Secondly, the amounts of DHEA and its sulfate administered in many of the *in vivo* and *in vitro* experiments, are what I term 'industrial size' doses, unknown to the human being. Would you like to comment as to whether we are seeing in these studies effects that might also be observed if you gave other steroids at that kind of dosage?

Vermeulen: I agree completely with the latter. In terms of physiology, a dose of 50 mg per 24 h given orally is probably acceptable. At this dosage you obtain plasma levels which are more or less comparable to levels we have in young adults. Taking into account the plasma levels of young men and a metabolic clearance rate of 15 l, you arrive at a blood production rate of DHEA and its sulfate of about 30–35 mg. The data of Morales and co-workers[2] relates to a dose of 50 mg. They found that the elderly people treated felt better in terms of well–being, although this subjective parameter is of course difficult to evaluate quantitatively. They also observed an increase in IGF–I. In elderly people, IGF–I levels are low, but I do not know whether this increase in IGF–I must be seen as being partly responsible for the increases in well-being or muscle mass. Doses of 300 and 1600 mg administered in other studies for short durations are certainly pharmacological doses. One wonders whether the effects observed wouldn't be obtained by administering just testosterone in small doses. We have to be very careful before claiming that DHEA has favorable effects, taking into account that these high doses probably have side-effects.

REFERENCES

1. Weksler, M. E. (1996). Hormone replacement for men. *Br. Med. J.,* **312**, 859–60
2. Morales, A. J., Nolan, J. J., Nelson, J. C. and Yen, S. S. C. (1994). Effects of replacement dose of dehydroepiandrostone in men and women of advancing age. *J. Clin. Endocrinol. Metab.* **78**, 1360–7

2

Epidemiology and diagnosis of osteoporosis in men

E. S. Orwoll

The earliest reports of the epidemiology of fractures associated with osteo-porosis revealed that the classical age-related increase in fractures seen in women is evident in men as well. Only in the last few years has it been recognized that the problem of osteoporosis in men represents an important public health issue, and that it also presents a unique array of scientific challenges and opportunities[1–3].

EPIDEMIOLOGY OF FRACTURES IN MEN

In women, the relationship between bone mineral density and fracture risk has become increasingly clear, and it is possible to discuss confidently the epidemiology of osteoporosis as defined either by the presence of atraumatic fractures or low bone mass[4,5]. In men there is less information available regarding the causation of fracture, and hence a discussion of osteoporosis epidemiology must currently be primarily related to fracture patterns.

The incidence of all fractures is higher in men than women early in life, probably as a result of serious trauma[6,7]. At about age 40–50 years there is a reversal of this trend, with fractures in general, but in particular those of the pelvis, humerus, forearm and femur becoming much more common in women. Nevertheless, the incidence of fractures due to minimal-to-moderate trauma (particularly hip and spine) also increases rapidly with aging in men (Figure 1)[8], and reflects an increasing prevalence of skeletal fragility.

Proximal femur

The proximal femur is the most important site of osteoporotic fracture, and the site about which the most complete epidemiological data are available.

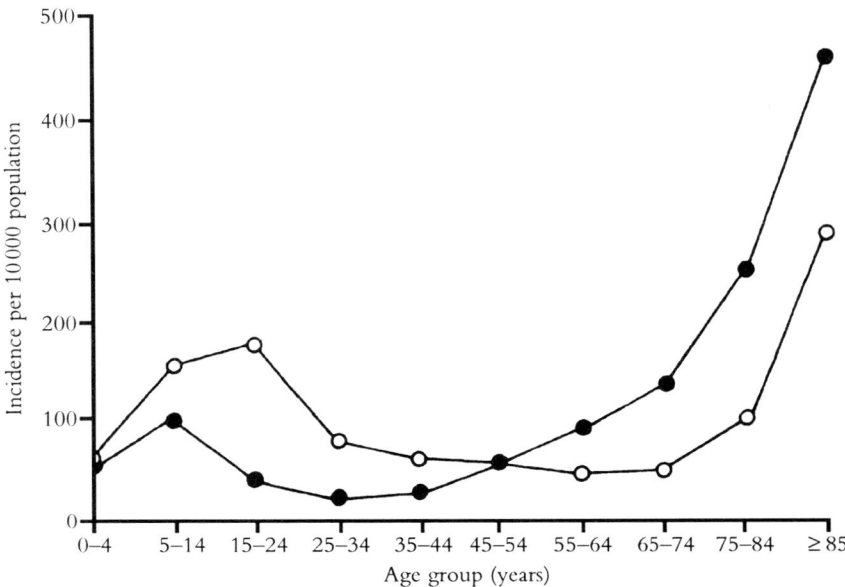

Figure 1 Average annual fracture incidence rate per 10 000 population in Leicester, UK, by age group and sex; open circles, males, solid circles, females (taken from ref. 8)

The incidence of hip fracture rises exponentially in men with aging, as it does in women. However, the age at which the increase begins is slightly older (approximately 5–10 years) in men[9]. In men from the USA older than 65 years, the incidence of hip fracture is 4–5/1000[10,11], compared with 8–10/1000 in similarly aged women in the USA. A similar 2 : 1 female : male ratio has been reported in northern Europe and Australia[12], although in other geographic areas the ratio has been noted to be much lower[13]. In southern Europe and other areas (e.g. Asia), the incidence of hip fracture is relatively lower in both sexes, and men have as many hip fractures as do women[14–17]. In Singapore the male : female incidence may actually be reversed[18]. Since there are fewer older men than women, the absolute number of hip fractures tends to be proportionally less in men (of those experiencing their first hip fracture at 65 years or older, 165 000 were men versus 580 000 women in the USA in 1984–87, or 22% of the fractures were in men)[11]. In the USA (Rochester) the lifetime risk of a hip fracture from age 50 onward has been calculated to be 6% in men and 17.5% in women[19], and 2.4% in men and 9% in women in Canada (Saskatchewan and Manitoba)[20]. Interestingly, the ratio of cervical to trochanteric hip fractures remains approximately 1.0 as age increases in men, whereas it decreases from 1.5 to 0.8 with aging in women[21].

Perhaps as a result of a higher prevalence of concomitant disease, the mortality associated with a hip fracture in elderly men (≥ 75 years) is considerably higher than in women (30% vs. 9%)[22,23]. In Europe the incidence of fracture is at least twice as great in women, but the death rates for femoral neck fractures are approximately equal, again suggesting a greater risk of mortality in men[24].

Unfortunately, the number of hip fractures is projected to increase dramatically as the elderly population expands[25]. Compounding matters, the age-adjusted incidence of hip fractures is also increasing in many areas. This increase in fracture incidence has been very consistent in the populations of men examined (primarily in the USA and northern Europe), whereas in women in the USA the rate of hip fracture appears to have stabilized[14,26,27]. It is estimated that approximately 30% of hip fractures world-wide will occur in men[28].

Racial differences in the incidence of hip fracture in men are substantial. For instance, African-American men experience hip fractures at a rate only half that of Caucasians[11]. Interestingly, whereas African-American women are at significantly lower risk for hip fracture than Caucasian women, African-American and Caucasian men are at similar risk, as are African-American women and men[9]. There are not extensive comparative data concerning other races, but one study clearly suggests a lower rate of hip fracture in Japanese compared to Caucasian men from the USA[29–31].

Bone mineral density has not been extensively examined in men with hip fracture, but as expected, density has been reported to be lower in subjects who suffer fractures[29,30]. Moreover, the proximal femoral densities associated with fracture in men tend to be higher than those similarly associated with fracture in women.

Vertebrae

Vertebral fracture is also an important sequel of osteoporosis. Since the diagnostic criteria for a vertebral fracture are unsettled, and vertebral fracture infrequently results in hospitalization, consistent epidemiological information is somewhat limited. Previously considered uncommon in men, recent information suggests that the incidence of osteoporotic vertebral fracture in US men is about half that in women (similar to hip fractures)[31–35]. In other studies, the prevalence of vertebral fracture is actually higher in men than women[36–38], and it is assumed that this represents an increase in the occurrence of early life trauma in men[34,39]. Fractures are primarily in low thoracic vertebrae in men, but are found at all levels. Most fractures are anterior compression in type[32] with vertebral crush fractures occurring less frequently than in women. Vertebral epiphysitis (Scheuermann's disease) is an uncommon cause of significant vertebral deformity in men[32].

Other fractures

Other fractures (radius/ulna, humerus, pelvis, femoral shaft) share a common epidemiological pattern. Men experience more of these fractures in youth, but with unusual exceptions (e.g. humerus) the incidence remains relatively stable during mid-life, while rising markedly in women[26,40,41]. It is only later (> 75 years) that the incidence of limb fracture begins to rise in men, and it then does so rapidly[40]. This increase is due primarily to an increase in lower limb fractures, while upper extremity fractures do not change as much. In older men, as in women, the likelihood of underlying bone pathology or propensity to fall (e.g. alcoholism) increased markedly[40]. Importantly, the occurrence of a distal forearm fracture[42] or a tibial fracture[43] in a man indicates a considerably increased risk of subsequent hip fracture, presumably as a result of low bone mass and/or an increased risk of falling.

DETERMINANTS OF FRACTURE IN MEN

Bone mass

The pattern of osteoporotic fractures is intimately related to aging, and changes in bone mass with age clearly contribute to that association. The decline in axial bone density was initially considered to be relatively slow in men, primarily because of cross-sectional studies using techniques that assess total spinal bone mass (dual photon absorptiometry [DPA]). Vertebral bone density as measured by quantitative computed tomography (QCT), however, suggested a much more rapid rate of bone loss with aging in normal men[44]. Subsequently, the results derived from DPA were shown to be influenced by artifacts in measurement introduced by extravertebral calcifications. If men with such calcifications are excluded, the relationship of spinal bone density to age is similar in men and women[32]. Longitudinal studies verify a more rapid rate of vertebral bone loss with aging in normal men[45]. In cross-sectional studies the negative slope of density with age at proximal femoral sites is similar, albeit somewhat less, in men compared to women (the menopausal acceleration of loss obviously does not occur in men)[46–48]. Moreover, bone volume in the iliac crest declines at very similar rates in both men and women.

In cortical bone the pattern of age-related loss also affects eventual fracture risk. Cross-sectional studies suggest that age is associated with a fairly linear decrease in cortical bone mass[45,46,49–52], but some also indicate the bone mineral density : age slope becomes more negative in men after 50 years[45,50,52,53]. This slope is not quite as steep as that in women[52], thereby accentuating the sexual differences in cortical mass present in early adulthood. However, the rate of loss of cortical bone mass in men as reported in longitudinal studies is considerably more rapid (5–10%/decade)[45,51,53,54] than previously estimated from cross-sectional studies (1–3%/decade)[46,47,50]. The differences noted in longitudinal *versus* cross-sectional studies may reflect the difficulty in adequately estimating time-dependent processes by cross-sectional methods, but also suggest that an increasingly greater rate of bone loss in men has taken place over the last several decades. This possibility is in accord with an apparent increase in fracture incidence (see above)[26].

There are few data available in men, but those available are consistent with an inverse relationship of bone mass to fracture. For instance, in men with spinal fractures assessments of femoral cortical area and Singh grade[55], proximal femoral DPA[32], vertebral QCT[56,57], vertebral DPA[45,58] and total body dual energy X-ray absorptiometry (DXA)[59] have all revealed reduced mean values compared with control men. Whereas in women it is suggested that levels of spinal bone density > 2 standard deviations below the mean of young normal subjects can be considered abnormal[19] (> 90% of patients with vertebral fractures have vertebral densities below this level), there is not enough experience to determine whether this is a reasonable approach in men as well. As in women, there is a clear overlap of bone density in men with fractures and non-fracture control subjects, indicating that bone density is not the sole determinant of vertebral fracture risk. Fracture is a somewhat chance event, and time as well as strength and propensity to fall are also important variables[30]. There are few specific data concerning the measurement of bone mass in men with hip fracture, although several reports have recently observed that hip and spine bone mineral density are clearly reduced in a series of men with hip fracture when compared with age-matched controls[29,30,60]. In a prospective study, Gardsell and colleagues showed that forearm bone density measures at both proximal and distal sites are lower in men who go on to sustain osteoporotic fractures (vertebrae, hip, proximal humerus, forearm, pelvis and tibial condyle) in the subsequent 10-year study period[61]. Moreover, in the Dubbo Osteoporosis Epidemiology Study it was found that femoral bone density measures were quite predictive of subsequent atraumatic fractures (although spinal bone mineral density was not)[30]. Vertebral and femoral bone mineral density values are reduced in men with vertebral fractures compared with non-fractured controls[32,62], indicating that vertebral fracture in men is not merely the result of a higher rate of trauma, but is also related to the presence of low bone mass. Finally, the studies of bone density in men with vertebral or femoral fractures suggest that the average bone density in fractured men is somewhat higher than that previously reported in similar studies in women[30,63,64]. This finding argues that the relationship between bone density and fracture risk must be independently evaluated in men and women.

Bone architecture

There are changes in bone architecture with aging and osteoporosis that in all likelihood influence fracture risk. The decline in vertebral trabecular number and thickness with age is associated with a reduction in compressive strength[65], and men with vertebral and femoral fractures have a lower trabecular plate density[66]. In men and women there is a generalized loss of trabeculae, but loss of horizontal elements (number and thickness) is particularly marked, in turn resulting in less support to vertical, load-bearing trabeculae[67]. Similar changes in trabecular structure in other locations (e.g. proximal femur) probably also contribute to fracture risk. In fact the quantification of proximal femoral trabecular patterns reveals a definite loss of trabeculation with age in men, and men with osteoporotic fractures have less trabeculation than control men[68]. In addition to trabecular loss, the appearance of micro-fractures increases with age, and may also increase skeletal fragility[69].

In mechanical terms the decline in cortical bone mass noted above is to some extent compensated by changes in cortical dimensions. In both sexes, there is an age-related increase in cortical width, and since fracture resistance is so dependent on geometry, this change is beneficial. In a two-decade study Garn and co-workers found that the rate of metacarpal cortical loss was very similar in both men and women, but periosteal apposition was somewhat greater in men (and endocortical loss somewhat less), mitigating the loss of thickness and overall mass[52]. This gender difference can also be observed in other long bones[70]. Interestingly, although men of greater weight and lean body mass have larger appendicular cortical areas, this does not protect them from age-related loss[52]. The increased periosteal apposition rate and the somewhat lesser rate of cortical density change with age in men are in accord with the fracture patterns observed in the elderly, in whom the rate of appendicular fractures is less in men than women. On the other hand, men experience an increase in porosity in the cortical component of bone, although at a rate somewhat slower than that seen in women[71,72]. This results in a reduction in density, and presumably in mechanical strength[70] and probably increases fracture risk.

Falls

In addition to bone mass, the risk of falling has been identified as a major determinant of fracture in women. In men, there are few prospective data that directly relate fall propensity to subsequent fractures, but a variety of factors indirectly related to risk of falling are associated with fracture. For instance, Nguyen and colleagues found that men who had experienced a non-traumatic fracture exhibit more body sway and lower grip strength (as well as lower bone density) that non-fracture controls[30]. Similarly, in a study of men with hip fractures[73] a number of factors associated with falls were found to be more prevalent than in controls. These included neurological disease, confusion, 'ambulatory problems' and alcohol use. As in women, the use of several classes of psychotropic drugs is associated with hip fracture risk in men[74,75]. Finally, men with hip fracture are of lower weight, have lower fat and lean body mass, and more commonly live alone than control subjects[29]. These differences suggest a body habitus and lifestyle more conducive to falls and injury, as well as the possibility of other interacting risk factors (nutritional deficiencies, comorbidities).

THE RELATIONSHIP BETWEEN THE AGE-RELATED DECLINE IN ANDROGEN LEVELS AND BONE LOSS IN MEN

There is a well-documented fall in both testicular and adrenal androgen concentrations in men with aging. Clinical hypogonadism is a well-established cause of bone loss in men, and an important issue is whether the decline in androgen levels with aging contributes to the fall in bone mass (and hence the increase in fracture risk) that occurs concomitantly. A number of studies have examined the relationship between androgen levels and bone mass in non-hypogonadal men, but the results have been inconclusive. In some, there has been a correlation (albeit weak) between androgen levels and measures of bone density[76–78], but in other reports when age is considered in the statistical model it is not possible to document a clear influence of androgen levels on bone[79–81]. Unfortunately, these studies have utilized small numbers of subjects and have been cross-sectional in design. There have been few attempts to assess more directly the influence of androgens in elderly men. In one notable report, Tenover found that a 3-month period of testosterone supplementation in a small group of elderly men with low normal testosterone levels at baseline was associated with a significant decline in urinary hydroxyproline excretion (although there was no change in osteocalcin levels)[82]. In sum, the issue of whether declines in androgen activity with aging have a clear impact on skeletal health remains unresolved. Important questions include whether there is a threshold level of androgen activity which is necessary for the maintenance of skeletal health, whether some skeletal compartments are more affected than others by changes in androgen levels, whether androgens may affect the skeleton *via* indirect effects on other tissues (e.g. muscle or body composition), and whether androgen supplementation is capable of preventing or reversing age-related changes in skeletal mass.

THE DIAGNOSIS OF OSTEOPOROSIS IN MEN

Guidelines for the most efficient, cost-effective approach for the evaluation of the patient with low bone mass, or the patient suspected of having low bone mass, are poorly validated for either sex. Recommendations are therefore based on existing knowledge of disease epidemiology and clinical characteristics[83,84] rather than upon models that have been carefully tested in prospective studies. Within these constraints it is possible to formulate an approach to the male osteoporotic (Figure 2).

In some men the diagnosis of an osteopenic metabolic bone disease can be made with basic clinical information. Most important is a clear history of low trauma fractures in the absence of evidence of a focal pathological

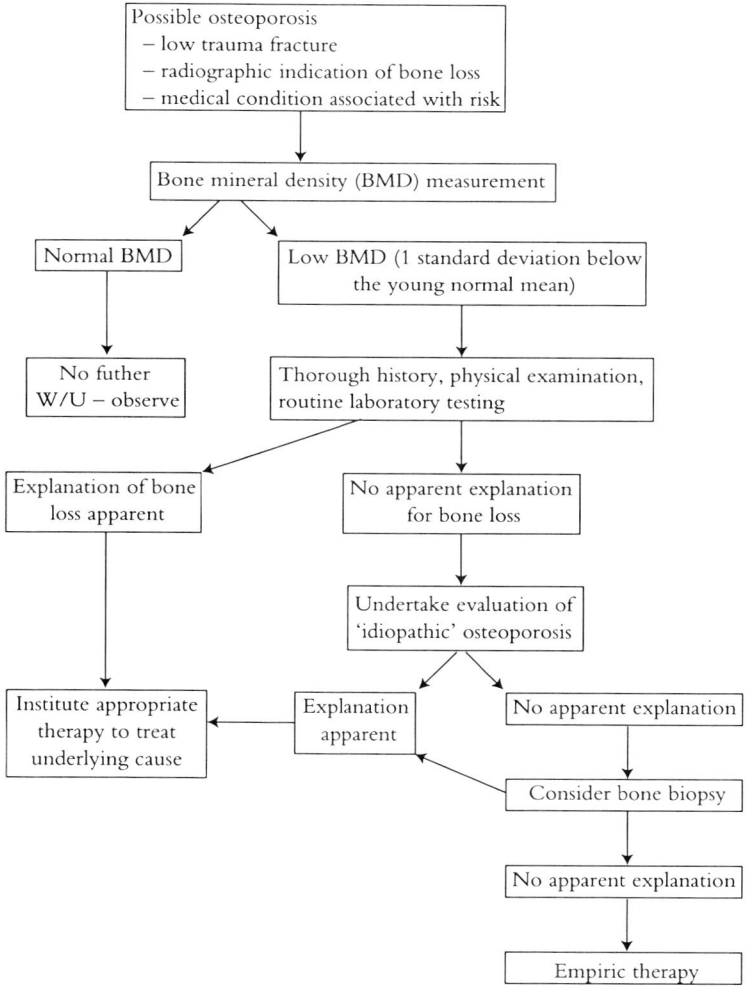

Figure 2   Schematic approach to the evaluation of osteoporosis in men[86]

process (malignancy, infection, Paget's disease). There are several clinical situations in which the presence of osteoporosis cannot be confidently determined, but should be considered likely. In these circumstances further diagnostic steps are appropriate. These situations include the presence of suspicious fractures, the radiographic presence of low bone mass, and conditions known to be associated with increased risk of bone loss.

Fractures

A wide variety of fractures[83,84] are associated with low bone mass in women, and thus can practically be termed osteoporotic[85]. In men there are fewer data, but the publications that have examined the relationships between bone mass, bone structure, mechanical strength and fracture in men support this contention[61,67]. Thus the presence of a low trauma fracture should raise the probability of metabolic bone disease and prompt further evaluation (i.e. densitometry). Certainly the occurrence of classic osteoporotic fractures (vertebral, proximal, femoral) in the absence of focal pathology should raise immediate concern. Even incidentally discovered vertebral deformities have been shown to be associated with lower bone mass[32], and should raise the suspicion of metabolic bone disease.

Radiographic detection of reduced bone mass

Low bone mass apparent radiographically, even in the absence of a fracture, is of concern because the loss of bone must be well advanced before it is detectable with routine X-ray procedures. It should be verified by a quantitative measure of bone mass.

Clinical conditions associated with low bone mass

There is a spectrum of causes of increased risk for osteoporosis in men, including glucocorticoid excess, alcoholism and hypogonadism and a wide variety of others[86,87] (Table 1). The presence of one, or particularly several, of these conditions should prompt concern, and the consideration for further characterization of the skeletal status.

Table 1 Causes of osteoporosis in men

I.	Primary
	Senile
	Idiopathic
II.	Secondary
	Hypogonadism
	Glucocorticoid excess
	Alcoholism
	Gastrointestinal disorders
	Hypercalciuria
	Smoking
	Anticonvulsants
	Thyrotoxicosis
	Immobilization
	Osteogenesis imperfecta
	Homocystinuria
	Systemic mastocytosis
	Neoplastic diseases
	Rheumatoid arthritis

Bone mineral density measurements

In men who present with findings that suggest the presence of metabolic bone disease (low trauma fractures, radiographic criteria indicating the presence of a reduction in bone mass, or conditions associated with bone loss), the measurement of bone mineral density should be strongly considered. The National Osteoporosis Foundation has suggested guidelines for the use of bone mineral density assessments that are in accordance with these recommendations[88,89]. Bone mineral density measurements can be useful in several ways, including cementing the diagnosis of low bone mass and gauging the severity of the process.

For the reasons that have been previously elaborated[88,89], bone density measures provide valuable data that cannot be deduced from other clinical information, and can solidify the diagnosis of low bone mass. Although this contention is derived from studies in women, the basic tenets should be applicable in men as well. Specifically,

(1) Low bone mass is related to fracture;

(2) Bone mass measures predict fracture risk;

(3) Bone mass can be accurately measured;

(4) An understanding of bone mass may influence the therapeutic approach; and

(5) Treatment of osteoporosis affects fracture risk.

There have been few prospective attempts to relate bone mineral density to eventual fracture risk in men, and hence the implications of a specific level of bone density are difficult to judge. The World Health Organization has proposed that in women *osteopenia* should be identified when density is less than 1.5 standard deviations below the young normal mean, and *osteoporosis* when the density is less than 2.5 standard deviations below[5]. Whether a similar approach can be taken in the diagnosis of osteopenia and osteoporosis in men is unknown. Looker and colleagues have recently attempted to apply the 1 and 2.5 standard deviation criteria derived from young normal male density levels to identify what proportion of the overall US male population would have osteopenia or osteoporosis[90]. They found that almost 50% of men aged 50 or more would have osteopenia, and 5% would have osteoporosis using these criteria. The usefulness of the application of these criteria is yet to be established.

Differential diagnosis

The intent at this stage of the evaluation should be to determine with reasonable certainty the histological cause of the osteopenic disorder, and to identify the etiologic factors contributing to it. In women, the vast majority of patients with osteopenic fractures have histological osteoporosis, but a small proportion are found to be osteomalacic[91–93]. Similarly, a fraction of men with fracture have osteomalacia[91–93]. Osteomalacia is estimated to be present in < 4% to 47% of men with femoral fractures, with most reports being ≤ 20%[91–95]. Since food is fortified with vitamin D, occult osteomalacia may be less frequent in the USA than in other areas (e.g. northern Europe). Increasing age is associated with an increasing prevalence of osteomalacia[93]. Thus far, the only patients who have been carefully surveyed are those with femoral fractures, and it is not known whether populations with other fractures (vertebral) would include similar proportions of osteoporotic and osteomalacic individuals. Some have suggested that women with femoral fracture are more frequently osteomalacic than men[91,92], but others report no distinction[93]. Although the exact magnitude of the problem presented by osteomalacia in men is uncertain, it is clear that any differential diagnosis of low bone mass and fractures in men must consider the possibility. This becomes particularly imperative because the treatment for osteomalacia differs considerably from that of osteoporosis[96].

Initial evaluation: history, physical and routine biochemical measures

The history, and the physical and routine biochemical profiles can be very helpful in directing a focused evaluation of a man with low bone mass. A variety of approaches for the differential diagnosis of low bone mass have been suggested using standard clinical and biochemical information[96–98]. The goals of this stage of the evaluation should be to determine the specific diagnosis (what is the cause of the low bone mass – osteoporosis or osteomalacia?), and to identify contributing factors in the genesis of the disorder. Of particular importance in the history and physical profile therefore, are clinical signs of genetic, nutritional/environmental, social (alcohol, tobacco), medical, or pharmacological factors that may be present to aid in these goals. Routine laboratory testing should include levels of serum creatinine, calcium, phosphorus, alkaline phosphatase, and liver function tests, as well as a complete blood count. If, on the basis of these tests, there is evidence for medical conditions associated with bone loss (alcoholism, hyperparathyroidism, malignancy, Cushing's syndrome, thyrotoxicosis, malabsorption, etc.) a definitive diagnosis should be pursued with appropriate testing.

Evaluation of the patient with 'idiopathic' osteoporosis

In men with reduced bone mass in whom no clear pathophysiology is identified by the routine methods above, it has been considered appropriate to be diagnostically aggressive, primarily because the potential for occult 'secondary' causes of osteoporosis may be higher in men. However, the incidence of occult causes of osteoporosis in men, or whether it is greater than in women, is poorly studied. The diagnostic yield and cost effectiveness of extensive biochemical studies in the man with apparently 'idiopathic' osteoporosis are unknown. Even when lacking this information, a reasonable evaluation of the man without a clear etiology for osteoporosis might include:

(1)   24-h urine calcium and creatinine, to identify idiopathic hypercalciuria;

(2)   24-h urine cortisol;

(3)   Serum calcium, phosphorus and alkaline phosphatase;

(4)   Serum 25-hydroxy-vitamin D level;

(5)   Serum testosterone;

(6)   Serum thyroid stimulating hormone level; and

(7)   Serum protein electrophoresis (in those > 50 years to exclude multiple myeloma).

Histomorphometric characterization

Transiliac bone biopsy is a safe and effective means of assessing skeletal histology and remodeling characteristics[99]. Some have suggested a transiliac bone biopsy is indicated in those men in whom a thorough biochemical evaluation has failed to reveal an etiology for osteoporosis[100]. The rationale for this approach is based on the need to accomplish several objectives:

(1)   Ensure that occult osteomalacia is not present;

(2)   Identify unusual causes of osteoporosis that may be revealed only by histological analysis, such as mastocytosis[101,102]; and

(3)   Yield information concerning the remodeling rate, which in turn may further direct the differential diagnosis (e.g. unappreciated thyrotoxicosis or secondary hyperparathyroidism suggested by the presence of increased turnover) or may be helpful in designing the most appropriate therapeutic approach.

Considerable histologic heterogeneity exists among men with osteoporosis. Whether distinct histologic patterns represent different stages of a single disease entity, separate subtypes of the disease, or simply an arbitrary subdivision of a normal distribution of remodeling rates is unknown.

A reasonable approach to the evaluation of remodeling dynamics in men with idiopathic osteoporosis (no etiology apparent from non-invasive testing) may be to combine the advantages of the biochemical markers of bone turnover with those of bone biopsy. An initial biochemical assessment of bone turnover should provide an understanding of remodeling rate. In the presence of an increase in biochemical indices of remodeling (osteocalcin, pyridinoline), a bone biopsy may be appropriate to identify unusual causes of low bone mass (e.g. mastocytosis). Bone biopsy may also be particularly helpful if there is any clinical concern for occult osteomalacia. Although alkaline phosphatase levels are usually increased in osteomalacia[96], even in this situation, a bone biopsy can reveal unanticipated osteomalacia, particularly in older men[93]. Unfortunately, the diagnostic yield or clinical impact of the bone biopsy is unknown. There is concern that it may be low, thus detracting from its clinical applicability.

REFERENCES

1. Niewoehner, C. (1993). Osteoporosis in men: is it more common than we think? *Postgrad. Med.*, **93**, 59–58

2. Seeman, E. (1993). Osteoporosis in men: epidemiology, pathophysiology, and treatment possibilities. *Am. J. Med.*, **95**, 22S–8S

3. Scane, A. C., Sutcliffe, A. M. and Francis, R. M. (1992). Osteoporosis in men. *Clin. Rheum.*, **7**, 589–601

4. Melton, L. J. I. (1995). How many women have osteoporosis now? *J. Bone Miner. Res.*, **10**, 175–7

5. WHO (1994). Assessment of fracture risk and its application to screening for postmenopausal osteoporosis (report of a WHO study group). (Geneva: World Health Organization)

6. Arneson, T. J., Melton, L. J. I., Lewallen, D. G. and O'Fallon, W. M. (1988). Epidemiology of diaphyseal and distal femoral fractures in Rochester, Minnesota, 1965–1984. *Clin. Orthop. Rel. Res.*, **234**, 188–94

7. Melton, L. J. I. and Cummings, S. R. (1987). Heterogeneity of age-related fractures: implications for epidemiology. *J. Bone Miner. Res.*, **2**, 321–31

8. Donaldson, L. J., Cook, A. and Thomson, R. G. (1990). Incidence of fractures in a geographically defined population. *J. Epidemiol. Community Health*, **44**, 241–5

9. Farmer, M. E., White, L. R., Brody, J. A. and Bailey, K. R. (1984). Race and sex differences in hip fracture incidence. *Am. J. Public Health*, **74**, 1374–80

10. Bacon, W. E., Smith, G. S. and Baker, S. P. (1989). Geographic variation in the occurrence of hip fractures among the elderly white US population. *Am. J. Public Health*, **79**, 1556–8

11. Jacobsen, S. J., Goldberg, J., Miles, T. P., Brody, J. A., Stiers, W. and Rimm, A. A. (1990). Hip fracture incidence among the old and very old: a population-based study of 745,435 cases. *Am. J. Public Health*, **80**, 871–3

12. Jones, G., Nguyen, T., Sambrook, P. N., Kelly, P. J., Gilbert, C. and Eisman, J. A. (1994). Symptomatic fracture incidence in elderly men and women: the Dubbo osteoporosis epidemiology study (DOES). *Osteoporosis Int*, **4**, 277–82

13. Gallagher, J. C., Melton, L. J. and Riggs, B. L. (1980). Epidemiology of fracture of the proximal femur in Rochester, MN. *Clin. Orthop.*, **150**, 163–71

14. Kanis, J. A. (1993). The incidence of hip fracture in Europe. *Osteoporosis Int.*, **1**, S10–15

15. Chalmers, J. and Ho, K. C. (1970). Geographical variations in senile osteoporosis: the association with physical activity. *J. Bone Joint Surg. Br.*, **52B**, 667–75

16. Solomon, L. (1968). Osteoporosis and fracture of the femoral neck in the South African Bantu. *J. Bone Joint Surg. Br.*, **50B**, 2–13

17. Elffors, I., Allander, E., Kanis, J. A., Gullberg, B., Johnell, O., Dequeker, J., Dilsen, G., Gennari, C., Lopes Vaz, A. A., Lyritis, G., Mazzuoli, G. F., Miravet, L., Passeri, M., Cano, R P., Rapado, A. and Ribot, C. (1994). The variable incidence of hip fracture in Southern Europe: the MEDOS study. *Osteoporosis Int.*, **4**, 253–63

18. Wong, P. C. N. (1966). Fracture epidemiology in a mixed southeastern Asian community (Singapore). *Clin. Orthop.*, **45**, 55–61

19. Melton, L. J. III and Chrischilles, E. A. (1992). Perspective: how many women have osteoporosis? *J. Bone Miner. Res.*, **7**, 1005–10

20. Martin, A. D., Silverthorn, K. G., Houston, C. S., Bernhardson, S., Wajda, A. and Roos, L. L. (1991). The incidence of fracture of the proximal femur in two million Canadians from 1972 to 1984. *Clin. Orthop.*, **266**, 111–18

21. Hinton, R. Y., Lennox, D. W., Ebert, F. R., Jacobsen, S. J. and Smith, G. S. (1995). Relative rates of fracture of the hip in the United States. *J. Bone Joint Surg. Am.*, **77A**, 695–702

22. Melton, L. J. III and Riggs, B. L. (1983). Epidemiology of age-related fractures. In *The Osteoporotic Syndrome*, Avioli, L. V. (ed.), pp. 45–72. (New York: Grune & Stratton)

23. Myers, A. H., Robinson, E. G., Van Natta, M. L., Michelson, J. D., Collins, K. and Baker, S. P. (1991). Hip fractures among the elderly: factors associated with in-hospital mortality. *Am. J. Epidemiol.*, **134**, 1128–37

24. Heyse, S. P. (1993). Epidemiology of hip fractures in the elderly: a cross-national analysis of mortality rates for femoral neck fractures. *Osteoporosis Int.*, **1**, S16–19

25. Schneider, E. L. and Guralnik, J. M. (1990). The aging of America: impact on health care costs. *J. Am. Med. Assoc.*, **263**, 2335–40

26. Melton, L. J. III, O'Fallon, W. M. and Riggs, B. L. (1987). Clinical investigations: secular trends in the incidence of hip fractures. *Calcif. Tiss. Int.*, **41**, 57–64

27. Falch, J. A., Kaastad, T. S. and Bohler, G. (1993). Secular increase and geographical differences in hip fracture incidence in Norway. *Bone*, **14**, 634–45

28. Cooper, C. and Campion, G. (1992). Hip fractures in the elderly: a world-wide projection. *Osteoporosis Int.*, **2**(6), 285–9

29. Karlsson, M. K., Johnell, O., Nilsson, B. E., Sernbo, I. and Obrant, K. J. (1993). Bone mineral mass in hip fracture patients. *Bone*, **14**, 161–5

30. Nguyen, T., Sambrook, P., Kelly, P., Jones, G., Lord, S. and Freund, J. (1993). Prediction of osteoporotic fractures by postural instability and bone density. *Br. Med. J.*, **307**, 1111–15

31. Ross, P. D., Norimatsu, H., Davis, J. W., Yano, K., Wasnich, R. D., Fujiwara, S., Hosoda, Y. and Melton, L. J. I. (1991). A comparison of hip fracture incidence among native Japanese, Japanese Americans, and American Caucasians. *Am. J. Epidemiol.*, **133**, 801–9

32. Mann, T., Oviatt, S. K., Wilson, D., Nelson, D. and Orwoll, E. S. (1992). Vertebral deformity in men. *J. Bone Miner. Res.*, **7**(11), 1259–65

33. Cooper, C., Atkinson, E. J., O'Fallon, W. M. and Melton, L. J. (1992). Incidence of clinically diagnosed vertebral fractures: a population-based study in Rochester, Minnesota, 1985–1989. *J. Bone Miner. Res.*, **7**, 221–7

34. Kanis, J. A. and McCloskey, E. V. (1992). Epidemiology of vertebral osteoporosis. *Bone*, **13**, S1–10

35. Cooper, C. and Melton, L. J. I. (1992). Epidemiology of osteoporosis. *Trends Endocrinol. Metab.*, **3**, 224–9

36. Davies, K. M., Stegman, M. R. and Recker, R. R. (1993). Preliminary vertebral deformity analysis for a rural population of older men and women. *J. Bone Miner. Res.*, **8**, S331

37. O'Neill, T. W., Valow, J. and Cooper, C. (1993). Differences in vertebral deformity indices between 3 European populations. *J. Bone Miner. Res.*, **8**, S149

38. Santavirta, S., Konttinen, Y. T., Heliovaara, M., Knekt, P., Luthje, P. and Aromaa, A. (1992). Determinants of osteoporotic thoracic vertebral fracture. *Acta Orthop. Scand.*, **63**(2), 198–202

39. Silman, A. J., O'Neill, T. W., Varlow, J. R., Agnusdei, D., Cooper, C., Dequeker, J., Felsenberg, D., Kanis, J. A., Kruskemper, G. and Raspe, H. (1995). Effect of physical activity on the risk of vertebral deformity. 17th Annual Meeting of the American Society of Bone and Mineral Research, Baltimore, MD, S174 (Cambridge MA: Blackwell Science Inc.)

40. Garraway, W. M., Stauffer, R. N., Kurland, L. T. and O'Fallon, W. M. (1979). Limb fractures in a defined population. I. Frequency and distribution. *Mayo Clin. Proc.*, **54**, 701–7

41. Buhr, A. J. and Cooke, A. M. (1959). Fracture patterns. *Lancet*, **I**, 531–6

42. Mallmin, H., Ljunghall, S., Persson, I., Naessen, T., Krusemo, U. B. and Bergstrom, R. (1993). Fracture of the distal forearm as a forecaster of subsequent hip fracture: a population-based cohort study with 24 years follow-up. *Calcif. Tissue Int.*, **52**, 269–72

43. Karlsson, M., Hasserius, R. and Obrant, E. J. (1993). Individuals who sustain nonosteoporotic fractures continue to also sustain fragility fractures. *Calcif. Tissue Int.*, **53**, 229–31

44. Meier, D. E., Orwoll, E. S. and Jones, J. M. (1984). Marked disparity between trabecular and cortical bone loss with age in healthy men: measurement by vertebral computed tomography and radial photon absorptiometry. *Ann. Intern. Med.*, **101**, 605–12

45. Orwoll, E. S., Oviatt, S. K., McClung, M. R., Deftos, L. J. and Sexton, G. (1990). The rate of bone mineral loss in normal men and the effects of calcium and cholecalciferol supplementation. *Ann. Int. Med.*, **112**, 29–34

46. Hannan, M. T., Felson, D. T. and Anderson, J. J. (1992). Bone mineral density in elderly men and women: results from the Framingham osteoporosis study. *J. Bone Miner. Res.*, **7**, 547–53

47. Riggs, B. L., Wahner, H. W., Dunn, W. L., Mazess, R. B., Offord, K. P. and Melton, L. J. I. (1981). Differential changes in bone mineral density of the appendicular and axial skeleton with aging. *J. Clin. Invest.*, **67**, 328–35

48. Elliott, J. R., Gilchrist, N. L., Wells, J. E., Turner, J. G., Ayling, E., Gillespie, W. J., Sainsbury, R., Hornblow, A. and Donald, R. A. (1990). Effects

of age and sex on bone density at the hip and spine in a normal caucasian New Zealand population. *NZ Med. J.*, **103**, 33–7

49. Gotfredsen, A., Hadberg, A., Nilas, L. and Christiansen, C. (1987). Total body bone mineral in healthy adults. *J. Lab. Clin. Med.*, **110**, 362–8

50. Mazess, R. B., Barden, H. S., Drinka, P. J., Bauwens, S. F., Orwoll, E. S. and Bell, N. H. (1990). Influence of age and body weight on spine and femur bone mineral density in U.S. white men. *J. Bone Miner. Res.*, **5**, 645–52

51. Davis, J. W., Ross, P. D., Vogel, J. M. and Wasnich, R. D. (1991). Age-related changes in bone mass among Japanese-American men. *Bone and Mineral*, **15**, 227–36

52. Garn, S. M., Sullivan, T. V., Decker, S. A., Larkin, F. A. and Hawthorne, V. M. (1992). Continuing bone expansion and increasing bone loss over a two-decade period in men and women from a total community sample. *Am. J. Hum. Biol.*, **4**, 57–67

53. Tobin, J. D., Fox, K. M. and Cejku, M. L. (1993). Bone density changes in normal men: a 4–19 year longitudinal study. *J. Bone Miner. Res.*, **8**, 102

54. Slemenda, C. W., Christian, J. C., Reed, T., Reister, T. K., Williams, C. J. and Johnston, C. C. J. (1992). Long-term bone loss in men: effects of genetic and environmental factors. *Ann. Intern. Med.*, **117**, 286–91

55. Francis, R. M., Peacock, M., Marshall, D. H., Horsman, A. and Aaron, J. E. (1989). Spinal osteoporosis in men. *Bone and Mineral*, **5**, 347–57

56. Genant, H. K., Gordan, G. S. and Hoffman, P. G. J. (1983). Osteoporosis Part I. Advanced radiologic assessment using quantitative computed tomography – Medical Staff conference, University of California, San Francisco. *West J. Med.*, **139**, 75–84

57. Resch, A., Schneider, B., Bernecker, P., Battmann, A., Wergedal, J., Willvonseder, R. and Resch, H. (1995). Risk of vertebral fractures in men: relationship to mineral density of the vertebral body. *Am. J. Radiol.*, **164**, 1447–50

58. Riggs, B. L., Wahner, H. W., Seeman, E., Offord, K. P., Dunn, W. L., Mazess, R. B., Johnson, K. A. and Melton, L. J. (1982). Changes in bone mineral density of the proximal femur and spine with aging. *J. Clin. Invest.*, **70**, 716–23

59. Hamdy, R. C., Moore, S., Cancellaro, V., Whalen, K., Landy, C., Chi, D., Reynolds, S. and Cambron, G. (1992). Osteoporosis in men – prevalence and biochemical parameters. 1992 Program and Abstracts of the 14th Annual Meeting of the American Society for Bone and Mineral Research, Minneapolis, MN, S325 #931 (New York: Mary Ann Liebert Inc.)

60. Chevalley, T., Rizzoli, R., Nydegger, V., Slosman, D., Tkatch, L., Rapin, C.-H., Vasey, H. and Bonjove, J-P. (1991). Preferential low bone mineral density of the femoral neck in patients with a recent fracture of the proximal femur. *Osteoporosis Int.* **1**, 147–54

61. Gardsell, P., Johnell, O. and Nilsson, B. E. (1990). The predictive value of forearm bone mineral content measurements in men. *Bone*, **11**, 229–32

62. Resch, H., Pietschmann, P., Woloszczuk, W., Krexner, E., Bernecker, P. and Willvonseder, R. (1992). Bone mass and biochemical parameters of bone metabolism in men with spinal osteoporosis. *Eur. J. Clin. Invest.*, **22**, 542–5

63. Mann, D. R., Rudman, C. G., Akinbami, M. A. and Gould, K. G. (1992). Preservation of bone mass in hypogonadal female monkeys with recombinant human growth hormone administration. *J. Clin. Endocrinol. Metab.*, **74**, 1263–9

64. Greenspan, S. L., Myers, E. R., Maitland, L. A., Kido, T. H., Krasnow, M. B. and Hayes, W. C. (1994). Trochanteric bone mineral density is associated with type of hip fracture in the elderly. *J. Bone Miner. Res.*, **9**, 1889–94

65. Mosekilde, L., Viidik, A. and Mosekilde, L. (1985). Correlation between the compressive strength of iliac and vertebral trabecular bone in normal individuals. *Bone*, **6**, 291–5

66. Parfitt, A. M. and Mathews, H. E. (1983). Relationships between surface, volume, and thickness of iliac trabecular bone in aging and in osteoporosis. *J. Clin. Invest.*, **72**, 1396–409

67. Mosekilde, L. (1989). Sex differences in age-related loss of vertebral trabecular bone mass and structure – biomechanical consequences. *Bone*, **10**, 425–32

68. Singh, M., Riggs, B. L., Beabout, J. W. and Jowsey, J. (1972). Femoral trabecular-pattern index for evaluation of spinal osteoporosis. *Ann. Int. Med.*, **77**, 63–7

69. Frost, H. M. (1993). Suggested fundamental concepts in skeletal physiology. *Calcif. Tissue Int.*, **52**, 1–4

70. Martin, B. (1993). Aging and strength of bone as a structural material. *Calcif. Tissue Int.*, **53**, S34–40

71. Laval-Jeantet, A.-M., Bergot, C., Carroll, R. and Garcia-Schaefer, F. (1983). Cortical bone senescence and mineral bone density of the humerus. *Calcif. Tissue Int.*, **35**, 268–72

72. Brockstedt, H., Kassem, M. and Eriksen, E. F. (1993). Age- and sex-related changes in iliac cortical bone mass and remodeling. *Bone*, **14**, 681–91

73. Grisso, J. A., Chiu, G. Y., Maislin, G., Steinmann, W. C. and Portale, J. (1991). Risk factors for hip fractures in men: a preliminary study. *J. Bone Miner. Res.*, **6**, 865–8

74. Ray, W. A., Griffin, M. R., Schaffner, W., Baugh, D. K. and Melton, L. J. III (1987). Psychotropic drug use and the risk of hip fracture. *N. Engl. J. Med.*, **316**, 363–70

75. Ray, W. A., Griffin, M. R. and Downey, W. (1989). Benzodiazepines of long and short elimination half-life and the risk of hip fracture. *J. Am. Med. Assoc.*, **262**, 3303–7

76. Rudman, D., Drinka, P. J., Wilson, C. R., Mattson, D. E., Scherman, F., Cuisinier, M. C. and Schultz, S. (1994). Relations of endogenous anabolic hormones and physical activity to bone mineral density and lean body mass in elderly men. *Clin. Endocrinol.*, **40**, 653–61

77. McElduff, A., Wilkinson, M., Ward, P. and Posen, S. (1988). Forearm mineral content in normal men: relationship to weight, height and plasma testosterone concentrations. *Bone*, **9**, 281–3

78. Murphy, S., Khaw, K.-T., Cassidy, A. and Compston, J. E. (1993). Sex hormones and bone mineral density in elderly men. *Bone Miner.*, **20**, 133–40

79. Meier, D. E., Orwoll, E. S., Keenan, E. J. and Fagerstrom, R. M. (1987). Marked decline in trabecular bone mineral content in healthy men with age: lack of association with sex steroid levels. *J. Am. Geriatr. Soc.*, **35**, 189–97

80. Drinka, P. J., Olson, J., Bauwens, S., Voeks, S., Carlson, I. and Wilson, M. (1993). Lack of association between free testosterone and bone density separate from age in elderly males. *Calcif. Tissue Int.*, **52**, 67–9

81. Wishart, J. M., Need, A. G., Horowitz, M., Morris, H. A. and Nordin, B. E. C. (1995). Effect of age on bone density and bone turnover in men. *Clin. Endocrinol.*, **42**, 141–6

82. Tenover, J. S. (1992). Effects of testosterone supplementation in the aging male. *J. Clin. Endocrinol. Metab.*, **75**, 1092–8

83. Eastell, R. and Riggs, B. L. (1980). Diagnostic evaluation of osteoporosis. *Endocrinol. Metab. Clin. North Am.*, **17**, 547–71

84. Lane, J. M. and Vigorita, V. J. (1984). Osteoporosis. *Orthop. Clin. North Am.*, **15**, 711–28

85. Seeley, D. G., Browner, W. S., Nevitt, M. C., Genant, H. K., Scott, J. C. and Cummings, S. R. (1991). Which fractures are associated with low appendicular bone mass in elderly women? *Ann. Intern. Med.*, **115**, 837–42

86. Orwoll, E. S. and Klein, R. F. (1995). Osteoporosis in men. *Endocr. Rev.*, **16**, 87–116

87. Kelepouris, N., Harper, K. D., Gannon, F., Kaplan, F. S. and Haddad, J. G. (1995). Severe osteoporosis in men. *Ann. Intern. Med.*, **123**, 452–60

88. Johnston, C. C. J. (1989). Clinical indications for bone mass measurements: a report from the scientific advisory board on the national osteoporosis foundation. *J. Bone Miner. Res.*, **4**, 1–28

89. Johnston, C. C. Jr, Slemenda, C. W. and Melton, L. J. III (1991). Clinical use of bone densitometry. *N. Engl. J. Med.*, **324**, 1105–9

90. Looker, A. C., Johnston, C. C., Wahner, H. W. J., Dunn, W. L., Calvo, M. S., Harris, T. B., Heyse, S. P. and Lindsay, R. L. (1995). Defining low femur bone density levels in men. 17th Annual Meeting of the American Society for Bone and Mineral Research, Baltimore, MD, S468 (Cambridge MA: Blackwell Science Inc.)

91. Campbell, G. A., Hosking, D. J., Kemm, J. R. and Boyd, R. V. (1984). How common is osteomalacia in the elderly? *Lancet*, **II**, 386–8

92. Aaron, J. E., Stasiak, L., Gallagher, J. C., Longton, E. B., Nicholson, M., Anderson, J. and Nordin, B. E. C. (1974). Frequency of osteomalacia and osteoporosis in fractures of the proximal femur. *Lancet*, **I**, 229–33

93. Hordon, L. D. and Peacock, M. (1990). Osteomalacia and osteoporosis in femoral neck fracture. *Bone Miner.*, **11**, 247–59

94. Sokoloff, L. (1978). Occult osteomalacia in American (USA) patients with fracture of the hip. *Am. J. Surg. Pathol.*, **2**, 21–30

95. Wilton, T. J., Hosking, D. J., Pawley, E., Stevens, A. and Harvey, L. (1987). Osteomalacia and femoral neck fractures in the elderly patient. *J. Bone Joint Surg.*, **69-B**, 388–90

96. Marel, G. M., McKenna, M. J. and Frame, B. (1986). Osteomalacia. In Peck, W. A. (ed.) *Bone and Mineral Res/4*, 4, pp. 335–413. (Amsterdam: Elsevier Science Publishers)

97. Johnson, B. E., Lucasey, B., Robinson, R. G. and Lukert, B. P. (1989). Contributing diagnoses in osteoporosis. *Arch. Intern. Med.*, **149**, 1069–72

98. Singer, F. R. (1987). Metabolic bone disease. In Felig, P., Baxter, J. D., Broadus, A. E., Frohman, L. A. (eds.) *Endocrinology and Metabolism*; pp. 1454–99 (New York: McGraw-Hill)

99. Klein, R. F. and Gunness, M. (1992). The transiliac bone biopsy: when to get it and how to interpret it. *Endocrinologist*, **2**, 158–68

100. Jackson, J. A. and Kleerekoper, M. (1990). Osteoporosis in men: diagnosis, pathophysiology, and prevention. *Medicine*, **69**, 137–52

101. Chines, A., Pacifici, R., Avioli, L. A., Korenblat, P. E. and Teitelbaum, S. L. (1993). Systemic mastocytosis and osteoporosis. *Osteoporosis Int.*, **1**, S147–9

102. Chines, A., Pacifici, R., Avioli, L. V., Teitelbaum, S. L. and Korenblat, P. E. (1991). Systemic mastocytosis presenting as osteoporosis: a clinical and histomorphometric study. *J. Clin. Endocrinol. Metab.*, **72**, 140–4

3

Androgens, bone metabolism and osteoporosis

J. M. Kaufman

Osteoporosis in men has only recently been recognized as a significant public health problem and literature data on the epidemiology, physiopathology and treatment of this condition are rather limited[1-5]. In men, aging is accompanied by a progressive decline of bone mass, a process which may be accentuated after the age of 50 years[5]. In fact, cross-sectional studies may have underestimated the rate of this age-related bone loss, as indicated by the findings of longitudinal studies, which also confirmed that bone loss persists and may even be accelerated in elderly men[6-9]. This bone loss results in an exponential increase in the incidence of fractures in aging in men, although at a somewhat more advanced age than is the case in women[5].

Estrogens are essential for the maintenance of adult bone mass in premenopausal women and it is generally accepted that sex hormone deprivation underlies at least part of the involutional skeletal changes occurring after menopause. In parallel, it is tempting to postulate that androgens may play an important role in the maintenance of skeletal integrity in men and this may, indeed, prove to be the case[5]. One may then ask the question whether the hypoandrogenism of elderly men[10] plays a significant role in the senile bone loss. Conversely, one may wonder why the on average high circulating sex hormone levels in elderly men, as compared to those in elderly women, are not able to prevent bone loss which occurs at rather similar rates in both sexes in the elderly[9].

The data available on the effects of androgens on the skeleton in health and disease are, however, rather limited in face of the complexity of the interactions of sex steroids with bone metabolism. Indeed, a comprehensive picture of the role of androgens in the skeleton should distinguish between effects of androgens on bone growth and bone remodeling. Furthermore, it should be recognized that androgens may have direct effects on bone, effects mediated through local humoral agents (cytokines, growth factors), indirect effects through modulation of the secretion and/or the action of calciotropic hormones, as well as other indirect effects such as on muscle mass or on renal handling of minerals. Assessment of the specific role of androgens in the skeleton is furthermore complicated by their interrelation with other sex steroids as far as production, metabolism and effects of these hormones are concerned. Finally, one should try to make a distinction between physiological and pharmacological effects of androgens in the skeleton.

The present brief overview will highlight only some of the better established facts concerning the role of androgens in bone metabolism as well as some important points of discussion in the literature and some areas in need of further research, with special attention to the possible role of androgens in the maintenance of skeletal integrity in elderly men.

ANDROGENS AND THE BUILD-UP OF PEAK BONE MASS

Growth and peak bone mass

In adult men, as is the case in women, the risk of osteoporotic fractures appears to be directly related to bone mass and bone density, as well as bone size[5]. Bone mass in adult men is the result of the bone mass acquired during childhood and adolescence and the cumulative bone loss that occurred thereafter. A deficient build-up of bone mass will, therefore, put a subject at risk for osteoporosis at a more advanced age.

The increase in mineralized bone mass during growth proceeds approximately in parallel with linear growth, this parallelism being most striking during the pubertal growth spurt[11-13]. Biochemical indices of osteoblast function, such as serum alkaline phosphatase activity and serum osteocalcin levels, closely reflect the growth velocity curve[12,14-17].

Over 90% of the adult bone mass has been acquired by the end of pubertal development. However, increase of mineralized bone mass continues at the time linear growth ceases and peak bone mass may be reached only several years after closure of the epiphyseal growth plates[18-20], although more recent studies have suggested that peak bone mass is achieved earlier than originally believed[13,20]. The increase of bone mass after completion of growth is explained by the lag time between bone formation and mineralization[11], by decreasing porosity and more extensive mineralization of previously formed bone, and by continued endosteal and periosteal apposition[18].

Role of androgens

Whereas there is no clear indication for a role of sex steroids, including those of adrenal origin, in the prepubertal growth phase, it is generally accepted that gonadal steroids are instrumental in the pubertal growth acceleration[21-23]. Several cross-sectional[11-14,17,19,24] and longitudinal[15,16] studies have shown a close temporal relationship between changes in serum testosterone levels or stage of pubertal development and linear growth, osteoblastic activity and/or accretion of mineralized bone mass during normal male pubertal development. These observations, as well as findings in hypogonadal boys before and during treatment with androgens[22,23,25,26], are consistent with the concept of a causal role of exposure to increased testosterone levels in the pubertal acceleration of linear growth and in accretion of mineralized bone mass. Notwithstanding a strong association between effects of androgens on linear growth and on accretion of mineralized bone mass during pubertal development, part of the androgen effects on build-up of bone mass appears to be dissociated from linear growth, as is suggested by the observed association of a low peak bone mass with normal adult height in men with hypogonadotropic hypogonadism[25,26] and in men with a history of delayed puberty[27].

Patients with hypogonadotropic hypogonadism present with osteopenia, affecting cortical as well as trabecular bone. Observations that osteopenia is present both before and after epiphyseal closure, and the lack of correlation between severity of osteopenia and duration of hypogonadism[25,26] are consistent with the view that osteoporosis in these patients results from deficient accretion of bone rather than from premature bone loss. Although correction of the hypogonadism in these men can increase bone mass, this effect is observed mainly in the patients with an immature skeleton and, in any case, many patients remain osteopenic[25,26,28].

The observations of a relatively low peak bone mass achieved by patients treated for hypogonadotropic hypogonadism and by subjects with a history of delayed puberty, suggest that the timing of exposure to androgens is important for optimal accretion of bone mass. In this regard, it is interesting to mention the findings indicating that the height gain during the growth spurt is reduced as a function of increasing age at puberty[23]. Indeed, if bone accretion during the growth spurt is essential for achievement of an optimal peak bone mass, then the occurrence of a limited growth spurt could affect the final peak bone mass, while an effect on final adult height is compensated for by the prolongation of the linear growth phase. In this regard, it is also interesting to note that a similar constellation of delayed skeletal maturation, tall stature and osteopenia was described in an adult male subject with estrogen resistance caused by a mutation in the gene encoding for the estrogen receptor[29].

Mechanisms of action of androgens

Several sites of action of androgens may be involved in their growth-promoting action in pubertal boys[22,23,30]. Indeed, androgens stimulate growth hormone (GH) secretion, act synergistically with the GH–insulin-like growth factor-I (IGF-I) axis to promote growth, and have limited direct growth-promoting effects in the absence of GH action.

The available data do not allow definite settlement of questions concerning the relative importance for growth and skeletal development of pure androgenic effects as opposed to effects mediated through aromatization of androgens to estrogens. Indeed, male subjects with androgen insensitivity achieve a quantitatively normal growth spurt[31] and in several studies growth acceleration in pubertal boys was better correlated to estrogen than to androgen levels. Nevertheless, the growth-promoting effects of androgens have also been obtained with non-aromatizable androgenic compounds[23]. In any case, the findings for a recently described male subject with resistance to estrogens seem to indicate that estrogens play a major role in skeletal maturation and in build-up and/or maintenance of bone mass[29].

Studies in male rats have shown that both androgen deficiency and androgen resistance (testicular feminization) affect bone modeling and consequently bone size, whereas cancellous bone density is affected only in androgen deficiency, suggesting that the slightly higher circulating estrogen levels in the androgen resistant rat are able to preserve cancellous bone mass in the absence of androgen activity[32,33]. The latter studies have also provided evidence indicating that androgens can exert direct effects on skeletal modeling, independently of the GH–IGF-I axis.

In any case, recent *in vitro* studies have confirmed that androgens can act directly on the bone. Specific androgen binding sites have been identified in human osteoblast-like cell lines[34-36] and androgens have been shown to stimulate osteoblast function as well as bone cell proliferation and differentiation[35,37,38].

ANDROGENS AND THE MAINTENANCE OF BONE MASS IN ADULT MEN

Male hypogonadism and osteoporosis

The occurrence of osteoporosis has been documented in hypogonadism of various origin, including idiopathic hypogonadotropic hypogonadism[25,26], hyperprolactinemic hypogonadism[39,40], Klinefelter's syndrome[41-44], anorexia nervosa[45], hemochromatosis[46] and other forms of hypogonadism of central or primary testicular origin[47]. Conversely, hypogonadism has been identified as a risk factor for spinal osteoporosis in studies of men with vertebral crush fractures[2,47-49] and as a possible risk factor for hip fractures in elderly men[50].

As has been discussed in an earlier section, in at least part of these hypogonadal men, such as in those with idiopathic hypogonadotropic hypogonadism[25,26], osteopenia is due principally to a deficient build-up of peak bone mass, rather than being the consequence of premature bone loss. However, osteoporosis has also been described in association with adult onset hypogonadism, in conditions such as hemochromatosis and hyperprolactinemia. The latter observations would then indicate that androgen deficiency in men, as is the case for estrogen deficiency in women, results in accelerated bone loss.

Hypoandrogenism and bone metabolism

Until recently, the bone loss effect of male hypoandrogenism was less firmly established than is now generally believed. Indeed, several of the reports on osteopenia in hypogonadal men do not allow to differentiate between effects of androgen deficiency on build-up of bone mass and on subsequent bone loss. Furthermore, other reports on osteoporosis in hypogonadism do not allow one to establish with certainty a causal relationship between the hypogonadism and the observed low bone mass, because of the existence of possible confounding factors linked to the underlying disease or congenital abnormality.

The majority of animal studies on the effects of androgen deficiency on bone mass have been performed in the rat. Castration in this species results in diminished cortical and trabecular bone mass in comparison with controls, although there have been some discordant findings[51]. The decrease in bone mass resulting from castration in old rats was shown to be preceded by increased bone turnover[52], and accompanied by increased bone resorption[53].

Longitudinal data in hypogonadal men are scarce. However, the effects of androgen deficiency on bone metabolism in adult men have been documented in a small group of otherwise healthy young adult men, who underwent orchiectomy[54]. These men presented with a high bone turnover, as apparent from increased values for biochemical indices of bone remodeling, and a rapid bone loss. Bone loss was positively correlated with urinary hydroxyproline excretion, and the values for the biochemical indices of bone turnover were shown to decrease under treatment with calcitonin. More recently, an increased bone turnover and a decrease in bone density were described in elderly men during a hypoandrogenic state induced by treatment with a long-acting gonadotropin releasing hormone (GnRH)-agonist for benign prostatic hyperplasia[55].

High values for biochemical indices of bone formation and bone resorption, indicating higher bone turnover, were also described in hypogonadal osteoporotic men when compared to eugonadal men with spinal osteoporosis[49]. Findings in patients with Klinefelter's syndrome[44] showed increased values for indices of bone resorption, but decreased values for bone formation (i.e. serum osteocalcin).

Histomorphometric data concerning hypogonadal men are rather limited and often conflicting. In a study by Delmas and Meunier[42] in men with Klinefelter's syndrome or hypogonadotropic hypogonadism, and in a study by Francis and colleagues[56] in hypogonadal men with vertebral crush fractures and low circulating levels of 1,25-dihydroxy-vitamin D, the major finding was depressed bone formation and mineralization, although at least some of the subjects studied had moderately increased resorption surfaces. These findings are clearly at variance with those in hypogonadal men with spinal osteoporosis studied by Jackson and co-workers[49] who described a moderately increased bone turnover with normal or moderately increased bone formation, the former being substantially suppressed in three out of four of these patients during testosterone treatment. The histomorphometric findings in the latter study are in agreement with biochemical evidence of increased bone turnover in the same patients as well as in orchiectomized men and in men with 'chemical castration' by treatment with a long-acting GnRH agonist[54,55]. Moreover, the finding of an increased bone turnover seems in accordance with observations that spinal osteoporosis in hypogonadal men is characterized by a loss of trabecular stuctures (loss of connectivity)[56], rather than by thinning of otherwise preserved trabeculae, as described in aging eugonadal men[57].

The lack of homogeneity in histomorphometric findings in hypogonadal men may reflect differences in duration of the hypogonadism, the severity of the hypoandrogenism as well as other differences such as in the vitamin D status. In fact, this variability of the histomorphometric findings is reminiscent of that seen in postmenopausal women with a history of vertebral crush fracture, who may present with decreased, normal or increased bone remodeling, the histomorphometric findings being representative only of the situation at the time of investigation[18].

In summary, the whole of the data seems to indicate that acquired hypoandrogenism in adulthood results in increased bone turnover and accelerated bone loss, as is the case in acquired female hypogonadism. There may be an additional defect in osteoblastic function, which may come to the fore after long duration of hypoandrogenism, but this, again, is not specific for the situation in male hypoandrogenism and has also been proposed as a contributing factor in the pathophysiology of postmenopausal osteoporosis[18].

There is clearly a need for more studies on the skeletal consequences of male hypoandrogenism as many questions remain unanswered. Most studies failed to show a clear relation between prevailing androgen levels and bone mineral status, possibly because single measurements of androgen levels may not be representative of long-term androgen impregnation. Furthermore, it is presently not known if there are threshold androgen levels below which the deleterious effects on the skeleton are expressed. Severe isolated hypoandrogenism in men is not a common condition and it would, therefore, be important to understand if and how relative hypoandrogenism might be a contributing factor in the pathophysiology of other forms of osteoporosis in men. In this regard, areas of special interest are the possible adverse skeletal effects of partial hypoandrogenism in aging men, in men treated with glucocorticoids[58] and in chronic alcoholism[59].

Mechanisms of action of androgens

From the foregoing discussion it is clear that androgens play a central role in the build-up and the maintenance of bone mass in men. Several possible direct and indirect mechanisms of action of androgens may contribute to these skeletal effects, but their relative importance for bone physiology and pathology remains largely undefined.

Estrogen-mediated versus intrinsic androgen effects

The designs of the available human studies do not allow us to distinguish intrinsic androgen actions from estrogen-mediated effects following aromatization of androgens. Indirect evidence suggests, however, that only part of the androgen effects are estrogen-dependent. Indeed, the circulating estrogen levels in adult men are lower than those required for maintenance of bone mass in women. Conversely, in women bone mass can be maintained in the presence of excess androgen levels despite the presence of a marked hypoestrogenism[60]. Furthermore, in a pharmacological setting, non-aromatizable androgenic compounds have clear effects on bone metabolism[51]. In castrated aged rats, bone loss can be prevented by treatment with non-aromatizable androgens[52].

As mentioned before, osteoblast-like human cells have been shown to contain specific binding sites for androgens and androgen actions on bone cells have been demonstrated *in vitro*[34–38].

There is, on the other hand, considerable evidence suggesting that at least part of the androgen effects on the bone may be mediated by aromatization to estrogens. Bone loss in castrated aged male rats can be prevented by administration of estrogens[52] and, along the same lines, pharmacological doses of estrogens have been shown to be able to maintain bone mass in hypoandrogenic adult men[61]. Furthermore, bone mass is preserved in androgen-resistant rats who present with slightly elevated circulating estrogen levels[32]. Androgens may also facilitate indirectly the action of estrogens on the bone by lowering sex hormone binding globulin. Finally, one has also to consider the possibility that androgen actions may be mediated through local aromatization within the bone tissue[62].

While the foregoing discussion focuses mainly on quantitative aspects of sex steroid actions, another way to approach the question of the intrinsic androgen effects on the bone is to look for qualitatively specific androgen effects. A main effect of estrogens on the adult skeleton seems to be a decrease of the level of bone remodeling in comparison with estrogen-deficient states[51,63]. Androgen stimulation of bone formation in the adult skeleton could possibly be regarded as a more androgen-specific action. However, evidence in support of the existence of such an action in the human is not yet conclusive. A frequently cited report by Baran and colleagues[64] pertains in fact to only two bone biopsies in a single male patient with hypogonadism of unspecified origin, whereas in the better documented study by Jackson and co-workers[49] bone formation was rather decreased during testosterone treatment. Although animal studies certainly provide evidence for a potential stimulatory action of androgens on bone formation[37,38], this effect is not necessarily specific for androgens and has also been described for estrogens[65].

In conclusion, androgen effects on the bone probably involve both intrinsic androgen effects and estrogen-mediated effects. Qualitatively, there is a rather striking similarity between physiological estrogen effects on bone in adult women and physiological androgen effects on bone in adult men. Further work is required to document better the specific androgen effects.

Direct versus indirect mechanisms of action

Although the direct mechanisms of action of androgens on bone remain to be fully elucidated, the available information does allow us to conclude that there is at least the potential for such a direct action on bone cells of androgens as such and/or of estrogens derived from aromatization of androgens. Both specific androgen receptors[34–36] and estrogen receptors[66] are present in osteoblast-like cells. Androgens have been shown to stimulate proliferation and activity of bone cells *in vitro*[35,37,38]. Furthermore, there have been reports of both 5α-reductase activity[67] and aromatase activity[62] in bone tissue.

Several possible indirect mechanisms of action of androgens on bone can be considered. Androgens may modulate the secretion and/or the action of calciotropic hormones. Calcitonin serum levels have been reported to be decreased in hypogonadal men[67,68] and testosterone administration has been shown to enhance the hypocalcemic response to calcitonin in castrated rats[69]. Androgens can modulate the response of human bone cells to parathyroid hormone[70].

There is some indication that androgens may influence renal 1α-hydro-xylation of 25-hydroxy-vitamin D in particular clinical situations. In a study in hypogonadal men, both low 1,25-dihydroxy-vitamin D levels and low fractional calcium absorption were corrected after testosterone replacement therapy[56]. An effect for 1α-hydroxylation was, however, not confirmed in other studies[26]. Possibly, androgen effects on 1,25-dihydroxy-vitamin D production may be apparent only in situations with low 25-hydroxy-vitamin D reserve.

Androgens may influence directly, or through interaction with other systemic hormones, the release and/or action of cytokines and growth factors. Androgens have been reported to increase osteoblast-stimulating activity of human breast cancer cells *in vitro*[71]. Effects of androgens to enhance mitogenesis and differentiation in bone cells may be mediated by induction of transforming growth factor-β and androgens may sensitize bone cells to the effects of fibroblast growth factor and IGF-II[38]. Concerning possible effects of androgens on the release of cytokines, it is noteworthy that androgens have modulatory effects on immune and inflammatory responses[72]. Production of the potent bone resorber interleukin-1 has been reported to be increased in a hypogonadal man[73].

Androgens may also influence bone metabolism through direct renal effects on calcium and phosphate handling[74]. Finally, another likely mechanism of action of androgens is through its positive effect on the nitrogen balance and muscle mass[72] which, in turn, will influence mechanical loading of the skeleton.

ANDROGEN THERAPY IN HYPOGONADAL MEN

Androgen therapy has been shown to prevent the deleterious effects of orchiectomy on the skeleton in the rat[33,52] and it appears that both aromatizable and non-aromatizable androgens are effective in this regard[52,75]. The effectiveness of androgen substitution therapy in the prevention of osteopenia associated with hypogonadism in men has not been studied systematically and even less so the effectiveness of such a therapy in the treatment of installed osteoporosis[1,5]. In fact, the available, mostly retrospective, literature data show that after several years of androgen substitution therapy, hypogonadal men still present with a clear deficit in bone mass[25,44,76]. This indicates that androgen substitution therapy, as applied in daily practice, is not fully effective, either because the patients present with a deficit in bone mass which is not (or no longer) reversible by androgen therapy or, alternatively, because the applied therapy regimens are inadequate.

In several studies, androgen replacement therapy in hypogonadal men resulted in limited increases in bone mineral density[26,28,77]. In men with idiopathic hypogonadotropic hypogonadism, androgen replacement resulted in an increase of cortical and trabecular bone mass in men with immature skeleton, while only a small increase in cortical bone was observed in men with fused epiphyses[26]. In men with hyperprolactinemic hypogonadism, correction of the hyperprolactinemia with normalization of androgen levels resulted in a significant increase of cortical bone mass with a small, non-significant, increase of vertebral bone mass[39]. Limited improvements of hypogonadism have been described in other forms of acquired hypogonadism[78].

The whole of the available data probably allow us to assume that androgen substitution has a beneficial effect on the evolution of bone mass in hypogonadal male patients, if not by significantly increasing bone mass, at least by preventing further accelerated bone loss. However, also this effect of androgens should be confirmed in long-term controlled clinical trials. Taking into account the effects of androgens on both the build-up and the maintenance of bone mass, it can be anticipated that such treatment will be most effective when started early, even though the rather limited available cross-sectional studies failed to show a relation between duration of hypogonadism and the magnitude of the deficit in bone mass. The findings of osteopenia in men with a history of delayed puberty suggest that the timing of androgen therapy may be of crucial importance[27].

Although androgens have occasionally been used in the treatment of osteoporosis in eugonadal men, i.e. those with androgen levels within the normal range, there is no single study which documents a beneficial effect and there is no clear theoretical background to support such practice. Although androgen treatment in clinical situations with relative hypoandrogenism, such as in glucocorticoid treatment, in chronic alcoholism and in aging, may be considered on theoretical grounds, reliable data on the usefulness of such a treatment are so far not available. The possible role of androgen deficiency in the development of senile osteopenia in men is considered in more detail in the next section.

ANDROGENS AND BONE MINERAL STATUS IN ELDERLY MEN

Leydig cell function declines in elderly men, which results in lower serum levels for total and free testosterone[10,79]. Whether this relative hypoadrogenism favors the development of senile osteopenia is an intriguing, but not fully explored possibility. Indices of bone mass have been reported to be positively correlated to androgen levels in elderly men[80]. However, as both androgen levels and bone mass are known to decline with age[5,10], it is rather difficult to establish a causal relationship with certainty.

Kelly and co-workers[81] observed in a relatively small group of men (subgroup of 23 in a total study population of 48 men), aged 21–79 years, a positive correlation of a free testosterone index with the bone mineral density at the forearm, but not at the spine and the proximal femur, after correction for age. Murphy and colleagues[82] found in a group of 134 men with a mean age of 69.5 years (65–76 years) a weak but significant positive correlation between a free androgen index and bone mineral density at the proximal femur after adjustment for the effects of body mass index and age; there was no significant correlation between adjusted androgen levels and spine bone mineral density. Rudman and co-workers[83] found that total testosterone was correlated with femoral neck bone mineral density in a group of men aged 58–95 years. However, no association, independent of age, was found between total or free androgen levels and bone mineral density at multiple skeletal sites in elderly men studied by Meier and colleagues[84] and Drinka and colleagues[85].

Wishart and colleagues[86] described a parallel decline in bone density and a free androgen index in men aged 20–83 years, but this association was no longer significant after correction for age.

We have studied a group of 57 healthy elderly men, participating in an ongoing population-based study in a semi-rural community. In this group, body mass index emerged as the strongest predictor of lumbar spine, forearm and total body bone mineral densities, as measured by dual photon X-ray absorptiometry. After correction for the effect of age and body mass index, total and free testosterone levels correlated weakly, but significantly, with forearm bone mineral density only (Table 1). It is interesting to note that the forearm is also the only skeletal site for which bone mineral density was positively correlated with lean body mass, whereas bone mineral density at other skeletal sites was significantly correlated with fat mass, as measured by

Table 1 Relation between testosterone levels and bone mineral density by dual photon X-ray absorptiometry at the lumbar spine (L2–L4), at the proximal femur and at the distal forearm: correlation coefficients partialized for age and body mass index

	log (Total testosterone)	*log* (Free testosterone)
L2–L4	0.03	0.02
Femur neck	−0.01	0.01
Femur trochanter	0.25	0.23
Total hip	0.09	0.15
Forearm		
ultradistal	0.36*	0.38*
mid	0.37*	0.23
proximal	0.32*	0.20
total region	0.38*	0.29

*$p < 0.05$

dual X-ray absorptiometry. (J. Kaufman and S. Goemaere, unpublished results).

Although the evidence for a relationship between age-adjusted androgen levels and bone mineral density in elderly men is rather sparse, this does not exclude that both age and androgen levels may be relevant to bone mineral density. Indeed, there are several possible explanations for the rather weak association between age-adjusted androgen levels and bone mineral density, which may include the failure of the androgen measurements to be representative for the long-term androgen status, the possible existence of threshold concentrations for the action of androgens on the bone, a lack of statistical power and, mainly, the cross-sectional rather than longitudinal design of the studies.

Presently no information is available on the long-term effects of androgen therapy on bone mineral density in elderly men. Limited data available on the effects of androgen treatment on biochemical indices of bone turnover in elderly men, have not been univocal. Morley and colleagues[87] described increased serum levels of osteocalcin, a marker of osteoblastic function, during androgen treatment; Tenover[88] described a reduced hydroxyprolinuria in a small group of elderly men treated with androgens; Orwoll and Oviatt[89] found no significant effect of androgen treatment on the biochemical indices of bone turnover in a larger group of elderly men.

In conclusion, more studies are needed in order to establish whether the relative hypoandrogenism of aging men plays a significant role in senile osteoporosis in men and, *a fortiori*, to evaluate the therapeutic potential of androgen therapy in elderly men. These studies should take into account the remarkably large interindividual variability in androgen levels in elderly men.

CONCLUSION

The possible mechanisms of the action of androgens on bone involve both pure androgen effects as well as estrogen–mediated effects, and direct effects on bone cells as well as a variety of indirect mechanisms of action. A large body of evidence supports the view that androgens play an important role in the pubertal acceleration of linear growth and accretion of mineralized bone mass in males, and part of the androgen effects on the build–up of peak bone mass are dissociated from the effects on linear growth. Adults with a history of temporary androgen deficiency during adolescence present with osteopenia as a consequence of deficient build–up of bone mass, as can be observed in men treated for idiopathic hypogonadotropic hypogonadism or with a history of delayed puberty.

There is evidence suggesting that part, but not all, of the androgen effects on the build–up of bone mass are mediated through estrogens, which play a key role in skeletal maturation. Substantial increases of bone mass during androgen treatment in hypogonadal men have been observed, but are largely restricted to men with a still immature skeleton. The whole of the data available indicate that male hypoandrogenism acquired in adulthood results in increased bone turnover and accelerated bone loss. However, a clear understanding of all skeletal consequences of acquired male hypoandrogenism will require more studies.

Although it seems reasonable to assume that adequate androgen substitution can prevent bone loss in hypogonadal men, systematic controlled studies on the effectiveness of androgen replacement therapy to prevent bone loss in hypogonadal men are lacking. Androgen replacement therapy may result in a limited increase of bone mass in some hypogonadal men with mature skeletons. However, retrospective studies indicate that in the setting of daily clinical practice, most hypogonadal men still present with a clear deficit in bone mass after several years of otherwise apparently adequate androgen replacement therapy.

There are no data indicating that androgen therapy may be useful in the treatment of eugonadal osteoporotic men. At present, controlled studies on the direct skeletal effects of androgen treatment in clinical situations characterized by a relative hypoandrogenism are still lacking. In this regard, there is a need for further clarification of the possible contributory role of relative hypoandrogenism in the development of male senile osteoporosis. Since the findings of several cross-sectional studies are not univocal and rather inconclusive, longitudinal studies are needed in order to clarify this important issue.

REFERENCES

1. Jackson, J. A. and Kleerekoper, M. (1990). Osteoporosis in men: diagnosis, pathophysiology, and prevention. *Medicine*, **69**, 137–52

2. Ringe, J. D. and Dorst, A. J. (1994). Osteoporose bei Männern. Pathogenese und klinische Einteilung bei 254 Fällen. *Dtsch. Med. Wschr.*, **119**, 943–7

3. Niewoehner, C. (1993). Osteoporosis in men: is it more common than we think? *Postgrad. Med.*, **93**, 58–9

4. Seeman, E. (1995). The dilemma of osteoporosis in men. *Am. J. Med.*, **98** (Suppl. 2A), 76S–88S

5. Orwoll, E. S. and Klein, R. F. (1995). Osteoporosis in men. *Endocr. Rev.*, **16**, 87–116

6. Orwoll, E. S., Oviatt, S. K., McClung, M. R., Deftos, L. J. and Sexton, G. (1990). The rate of bone mineral loss in normal men and the effects of calcium and cholecalciferol supplementation. *Ann. Intern. Med.*, **112**, 29–34

7. Davis, J. W., Ross, P. D., Vogel, J. M. and Wasnich, R. D. (1991). Age-related changes in bone mass among Japanese–American men. *Bone and Mineral*, **15**, 227–36

8. Slemenda, C. W., Christian, J. C., Reed, T., Reister, T. K., Williams, C. J. and Johnston, C. C. Jr. (1992). Long-term bone loss in men: effects of genetic and environmental factors. *Ann. Intern. Med.*, **117**, 286–91

9. Jones, G., Nguyen, T., Sambrook, P., Kelly, P. J. and Eisman, J. A. (1994). Progressive loss of bone in the femoral neck in elderly people: longitudinal findings from the Dubbo osteoporosis epidemiology study. *Br. Med. J.*, **309**, 691–5

10. Vermeulen, A. (1991). Clinical review 24: androgens in the aging male. *J. Clin. Endocrinol. Metab.*, **73**, 221–3

11. Krabbe, S., Christiansen, C., Rødbro, P. and Transbøl, I. (1979). Effect of puberty on bone growth and mineralisation. *Arch. Dis. Child.*, **54**, 950–3

12. Glastre, C., Braillon, P., David, L., Colhat, P., Meunier, P. J. and Delmas, P. D. (1990). Measurement of bone mineral content of the lumbar spine by dual energy X-ray absorptiometry in normal children: correlations with growth parameters. *J. Clin. Endocrinol. Metab.*, **70**, 1330–3

13. Bonjour, J. P., Theintz, G., Bucks, B., Slosman, D. and Rizzoli, R. (1991). Critical years and stages of puberty for spinal and femoral bone mass accumulation during adolescence. *J. Clin. Endocrinol. Metab.*, **73**, 555–63

14. Krabbe, S., Christiansen, C., Rødbro, P. and Transbøl, I. (1980). Pubertal growth as reflected by simultaneous changes in bone mineral content and serum alkaline phosphatase. *Acta Paediatr. Scand.*, **69**, 49–52

15. Krabbe, S. and Christiansen, C. (1984). Longitudinal study of calcium metabolism in male puberty. I. Bone mineral content, and serum levels of alkaline phosphatase, phosphate and calcium. *Acta Paediatr. Scand.*, **73**, 745–9

16. Krabbe, S., Hummer, L. and Christiansen, C. (1984). Longitudinal study of calcium metabolism in male puberty: II. Relationship between mineralisation and serum testosterone. *Acta Paediatr. Scand.*, **73**, 750–5

17. Johansen, J. S., Giwercman, A., Hartwell, D., Nielsen, C. T., Price, P. A., Christiansen, C. and Skakkebaek, N. E. (1988). Serum bone gla-protein as a marker of bone growth in children and adolescents: correlation with age, height, serum insulin-like growth factor I, and serum testosterone. *J. Clin. Endocrinol. Metab.*, **67**, 273–8

18. Parfitt, A. M. (1988). Bone remodeling: relationship to the amount and structure of bone and the pathogenesis and prevention of fractures. In Riggs, B. L. and Melton III, L. J. (eds.). *Osteoporosis: Etiology, Diagnosis and Management*, pp. 45–9. (New York: Raven Press)

19. Gilsanz, V., Gibbens, D. T., Roe, T. F., Carlson, M., Senac, M. O., Boechat, M. I., Huang, H. K., Schulze, E. E., Libanatic, C. R. and Cann, C. C. (1988). Vertebral bone density in children: effect of puberty. *Radiology*, **166**, 847–50

20. Ott, S. M. (1990). Attainment of peak bone mass (editorial). *J. Clin. Endocrinol. Metab.*, **71**, 1082A–1082C

21. Rechler, M. M., Nissley, S. P. and Roth, J. (1987). Hormonal regulation of human growth. *N. Engl. J. Med.*, **316**, 941–2

22. Zachman, M. (1988). Interrelation of GH and sex steroids. In Frish, H. and Laron, Z. (eds.). *Induction of puberty in hypopituitarism* pp. 25–36. Serono Symposia Review **6**: Rome

23. Bourguignon, J. P. (1988). Linear growth as a function of age at onset of puberty and sex steroid dosage: therapeutic implications. *Endocr. Rev.*, **9**, 467–88

24. Riis, B. J., Krabbe, S., Christiansen, C., Catherwood, B. D. and Deftos, L. J. (1985). Bone turnover in male puberty: a longitudinal study. *Calcif. Tissue Int.*, **37**, 213–17

25. Finkelstein, J. S., Klibanski, A., Neer, R. M., Greenspan, S. L., Rosenthal, D. I. and Crowley, W. F. Jr. (1987). Osteoporosis in men with idiopathic hypogonadotropic hypogonadism. *Ann. Intern. Med.*, **106**, 354–61

26. Finkelstein, J. S., Klibanski, A., Neer, R. M., Doppelt, S. H., Rosenthal, D. I., Segre, G. V. and Crowley, W. F. Jr. (1989). Increases in bone density during treatment of men with idiopathic hypogonadotropic hypogonadism. *J. Clin. Endocrinol. Metab.*, **69**, 776–83

27. Finkelstein, J. S., Neer, R. M., Biller, B. M. K., Crawford, J. D. and Klibanski, A. (1992). Osteopenia in men with a history of delayed puberty. *N. Engl. J. Med.*, **326**, 600–4

28. Devogelaer, J. P., De Cooman, S. and Nagant de Deux Chaisnes, C. (1992). Low bone mass in hypogonadal males. Effect of testosterone substitution therapy, a densitometric study. *Maturitas*, **15**, 17–23

29. Smith, E. P., Boyd, J., Frank, G. R., Takahashi, H., Cohen, R. M., Specker, B., Williams, T. C., Lubahn, D. B. and Korach, K. S. (1994). Estrogen resistance caused by mutation in the estrogen–receptor gene. *N. Engl. J. Med.*, **331**, 1056–61

30. Raisz, L. G. and Kream, B. E. (1981). Hormonal control of skeletal growth. *Ann. Rev. Physiol.*, **43**, 225–38

31. Zachman, M., Prader, A., Sobel, E. H., Crigler, J. F., Ritzen, E. M., Atarés, M. and Ferrandez, A. (1986). Pubertal growth in patients with androgen insensitivity. *J. Pediatr.*, **108**, 694–7

32. Vanderschueren, D., Van Herck, E., Suiker, A. M. H., Visser, W. J., Schot, L. P. C., Chung, K., Lucas, R. S., Einhorn, T. A. and Bouillon, R. (1993). Bone metabolism in the androgen resistant (testicular feminized) male rat. *J. Bone Miner. Res.*, **8**, 801–9

33. Vanderschueren, D. (1994). *Skeletal effects of androgen deficiency, androgen replacement and androgen resistance*, PhD Thesis, s.n. Leuven, Belgium. KUL libris 822712

34. Colvard, D., Eriksen, E. F., Keeting, P. E., Wilson, E. M., Lubahn, D. B., French, F. S., Riggs, B L. and Spelberg, T. S. (1989). Identification of

androgen receptors in normal human osteoblast-like cells. *Proc. Natl. Acad. Sci., USA*, **86**, 854–7

35. Benz, D. J., Haussler, M. R., Thomas, M. A., Speelman, B. and Komm, B. S. (1991). High affinity androgen binding and androgenic regulation of alpha 1 (I) – procollagen and transforming growth factor – beta, steady state messenger ribonucleic acid levels in human osteoblast-like osteosarcoma cells. *Endocrinology*, **128**, 2723–30

36. Orwoll, E. S., Otribrska, L., Romsey, E. E. and Keenan, E. J. (1991). Androgen receptors in osteoblast-like cell lines. *Calcif. Tissue Int.*, **49**, 182–7

37. Kasperk, Ch., Wergedal, J. E., Farley, J. K., Linkhart, T. A., Turner, R. T. and Baylink, D. J. (1989). Androgens directly stimulate proliferation of bone cells *in vitro*. *Endocrinology*, **124**, 1576–8

38. Kasperk, Ch., Fitzsimmons, R., Strong, D., Mohan, S., Jennings, J., Wegerdal, J. and Baylink D. (1990). Studies on the mechanism by which androgens enhance mitogenesis and differentiation in bone cells. *J. Clin. Endocrinol. Metab.*, **71**, 1322–9

39. Greenspan, S. L., Neer, R. M., Ridgway, E. C. and Klibanski, A. (1986). Osteoporosis in men with hyperprolactinemic hypogonadism. *Ann. Intern. Med.*, **104**, 777–82

40. Greenspan, S. L., Oppenheim, D. S. and Klibanski, A. (1989). Importance of gonadal steroids to bone mass in men with hyperprolactinemic hypogonadism. *Ann. Intern. Med.*, **110**, 526–31

41. Smith, D. A. S. and Walker, M. S. (1977). Changes in plasma steroids and bone density in Klinefelter's syndrome. *Calcif. Ris. Res.*, **22** (suppl.), 225–9

42. Delmas, P. and Meunier, P. J. (1981). L'ostéoporose au cours du syndrome de Klinefelter. *Nouv. Press Med.*, **10**, 687–90

43. Foresta, C., Ruzza, G., Mioni, R., Meneghello, A. and Baccichetti, C. (1983). Testosterone and bone loss in Klinefelter syndrome. *Horm. Metab. Res.*, **15**, 56–7

44. Horowitz, M., Wishart, J., O'Loughlin, P. D., Morris, H. A., Need, A. G. and Nordin, B. E. C. (1992). Osteoporosis and Klinefelters' syndrome. *Clin. Endocrinol. (Oxf.)*, **36**, 113–18

45. Rigotti, N. A., Neer, R. M. and Jameson, L. (1986). Osteopenia and bone fractures in a man with anorexia nervosa and hypogonadism. *J. Am. Med. Assoc.*, **256**, 385–8

46. Diamond, T., Stiel, D. and Pasen, S. (1989). Osteoporosis in hemochromatosis: iron excess, gonadal deficiency, or other factors. *Ann. Intern. Med.*, **110**, 430–6

47. Francis, R. M., Peacock, M., Marshall, D. H., Horsman, A. and Aaron, J. E. (1989). Spinal osteoporosis. *Bone and Mineral*, **5**, 347–57

48. Seeman, E., Melton, L. J. III, O'Fallong, W. M. and Riggs, B. L. (1983). Risk factors for spinal osteoporosis in men. *Am. J. Med.*, **75**, 977–83

49. Jackson, J. A., Kleerekoper, M., Parfitt, A. M., Rao, D. S., Villanueva, A. R. and Frame, B. (1987). Bone histomorphometry in hypogonadal and eugonadal men with spinal osteoporosis. *J. Clin. Endocrinol. Metab.*, **65**, 53–8

50. Stanley, H. L., Schmitt, B. P., Poses, R. M. and Deiss, W. P. (1991). Does hypogonadism contribute to the occurrence of a minimal trauma hip fracture in elderly men? *Am. J. Geriatr. Soc.*, **39**, 766–71

51. Schot, L. P. C. and Schuurs, A. H. W. M. (1990). Sex steroids and osteoporosis: effects of deficiencies and substitute treatments. *J. Steroid Biochem. Molec. Biol.*, **37**, 167–82

52. Vanderschueren, P., Van Herck, E., Suiker, A. M. H., Visser, W. J., Schot, L. P. C. and Bouillon, R. (1992). Bone and mineral metabolism in aged male rats: short and long term effects of androgen deficiency. *Endocrinology*, **130**, 2906–16

53. Danielsen, C. C., Mosekilde, L. and Andreassen, R. R. (1992). Long-term effect of orchidectomy on cortical bone from rat femur: bone mass and mechanical properties. *Calcif. Tissue Int.*, **50**, 169–74

54. Stepan, J. J., Lachman, M., Zverina, J., Pacovsky, V. and Baylink, D. J. (1989). Castrated men exhibit bone loss: effect of calcitonin treatment on biochemical indices of bone remodeling. *J. Clin. Endocrinol. Metab.*, **69**, 523–7

55. Goldray, D., Weisman, Y., Jaccard, N., Merdler, C., Chen, J. and Matzkin, H. (1993). Decreased bone density in elderly men treated with the gonadotropin – releasing hormone agonist decapeptyl (D-Trp[6]-GnRH). *J. Clin. Endocrinol. Metab.*, **76**, 288–90

56. Francis, R. M., Peacock, M., Aaron, J. E., Selby, P. L., Taylor, G. A., Thompson, J., Marshall, D. H. and Horsman, A. (1986). Osteoporosis in hypogonadal men: role of decreased plasma 1,25-dihydroxyvitamin D, calcium malabsorption, and low bone formation. *Bone*, **7**, 261–8

57. Parfitt, A. M., Mathew, C. H. E., Villanueva, A. R., Kleerekoper, M., Frame, B. and Rao, D. S. (1983). Relationship between surface, volume, and thickness of iliac trabecular bone in aging and in osteoporosis. *J. Clin. Invest.*, **72**, 1396–1409

58. MacAdams, M. R., White, R. H. and Chipps, B. E. (1986). Reduction of serum testosterone levels during chronic glucocorticoid therapy. *Ann. Intern. Med.*, **104**, 648–51

59. Valimaki, M., Salaspuro, M. and Ylikahri, R. (1982). Liver damage and sex hormones in chronic male alcoholics. *Clin. Endocrinol. (Oxf.)*, **17**, 469–77

60. Dixon, J. E., Rodin, A., Murby, B., Chapman, M. G. and Fogelman, I. (1989). Bone mass in hirsute women with androgen excess. *Clin. Endocrinol. (Oxf.)*, **30**, 271–7

61. Lips, P., Asscheman, H., Uitewaal, P., Netelenbos, J. C. and Gooren, L. (1989). The effect of cross-gender hormonal treatment on bone metabolism in male-to-female transsexuals. *J. Bone Miner. Res.*, **4**, 657–62

62. Schweikert, H. V., Wolf, L. and Romalo, G. (1995). Oestrogen formation from androstenedione in human bone. *Clin. Endocrinol. (Oxf.)*, **43**, 37–42

63. Christiansen, C. and Lindsay, R. (1990). Estrogens, bone loss and preservation. *Osteoporosis Int.*, **1**, 2–13

64. Baran, D. T., Bergfeld, M. A., Teitelbaum, S. L. and Avioli, L. V. (1978). Effect of testosterone therapy on bone formation in an osteoporotic hypogonadal male. *Calcif. Tissue Int.*, **26**, 103–6

65. Chow, J., Tobials, J. H. and Colston, K. W. (1992). Estrogen maintains trabecular bone volume in rats not only by suppression of bone resorption but also by stimulation of bone formation. *J. Clin. Invest.*, **89**, 74–8

66. Eriksen, E. F., Colvard, D. S., Borg, N. J., Graham, M. L., Mann, K. G., Spelsberg, T. C. and Riggs, B. L. (1988). Evidence of estrogen receptors in normal human osteoblast-like cells. *Science*, **241**, 84–8

67. Foresta, C., Busnardo, B., Ruzza, G., Zanalta, G. and Mioni, R. (1983). Lower calcitonin levels in young hypogonadic men with osteoporosis. *Horm. Metab. Res.*, **15**, 206–7

68. Foresta, C., Zanatta, G. P., Busnardo, B. and Scanelli, C. (1985). Testosterone and calcitonin plasma levels in hypogonadal osteoporotic young men. *J. Endocrinol. Invest.*, **8**, 377–9

69. Ogaba, E., Shimazawa, E., Suzuki, H., Yoshitoshi, Y., Asano, H. and Ando, H. (1970). Androgens and enhancement of hypocalcemic response to thyrocalcitonin in rats. *Endocrinology*, **87**, 421–6

70. Fukayama, S. and Tashjian, A. T. (1989). Direct modulation by androgens of the response of human bone cells (SaOs-2) to human parathyroid hormone (PTH) and PTH-related peptide. *Endocrinology*, **125**, 1789–94

71. Vicard, E., Valentin-Opran, A., Chenu, C., Delmas, P. D., Merinier, P. J. and Saez, S. (1986). Androgens increase osteoblast-stimulating activity of human breast cancer cells *in vitro. J. Steroid Biochem.*, **24**, 401–3

72. Mooradian, A. D., Morley, J. E. and Korenman, S. G. (1987). Biological actions of androgens. *Endocr. Rev.*, **8**, 1–28

73. Axelrod, D. W., Lachman, D. B., Judge, D. B., Mallette, L. E. and Gagel, R. F. (1989). Resorptive hypercalciuria and increased interleukin 1 in a young male with hypogonadism and osteoporosis: reversal with androgen treatment. *Clin. Res.*, **37**, 21A (abstract)

74. Spencer, H., Friedland, J. A. and Lewin, I. (1976). Effect of androgen on bone, calcium and phosphorus. In Kochakian, C. D. (ed.) *Anabolic–Androgenic Steroids. Handbuch der Experimentellen Pharmakologie*, pp. 419–40. (Berlin: Springer Verlag)

75. Turner, R. T. and Wakley, G. K. (1990). Differential effects of androgens on cortical bone histomorphometry in gonadectomized male and female rats. *J. Orthop. Res.*, **8**, 612–17

76. Wong, F. H. W., Pun, K. K. and Wang, C. (1993). Loss of bone mass in patient with Klinefelter's syndrome, despite sufficient testosterone replacement. *Osteoporosis Int.*, **7**, 281–7

77. Isaia, G., Mussetta, M., Pecchio, F., Sciolla, A., Di Stefano, M. and Molinatti, G. H. (1992). Effect of testosterone on bone in hypogonadal males. *Maturitas*, **15**, 47–51

78. Diamond, T., Stiel, D. and Posen, S. (1991). Effects of testosterone and venesection on spinal and peripheral bone mineral in six hypogonadal men with hemochromatosis. *J. Bone Miner. Res.*, **6**, 39–43

79. Vermeulen, A. and Kaufman, J. M. (1992). Editorial: Role of the hypothalamo-pituitary function in the hypoandrogenism of healthy aging. *J. Clin. Endocrinol. Metab.*, **75**, 704–6

80. Foresta, C., Ruzza, G., Mioni, R., Guarneri, G., Gribaldo, R., Meneghello, A. and Mastrogiacomo, I. (1984). Osteoporosis and decline of gonadal function in the elderly male. *Hormone Res.*, **19**, 18–22

81. Kelly, P. J., Pocock, N. A., Sambrook, P. N. and Eisman, J. A. (1990). Dietary calcium, sex hormones, and bone mineral density in men. *Br. Med. J.*, **300**, 1361–4

82. Murphy, S., Khaw, K. T., Cassidy, A. and Compston, J. E. (1993). Sex hormones and bone mineral density in elderly men. *J. Bone Miner. Res.*, **20**, 133–40

83. Rudman, D., Drinka, P. J., Wilson, C. R., Mattson, D. E., Scherman, F., Cuisinier, M. C. and Schultz, S. (1994). Relations of endogenous anabolic hormones and physical activity to bone mineral density and lean body mass in elderly men. *Clin. Endocrinol.*, **40**, 653–61

84. Meier, D. E., Orwoll, E. S., Keenan, E. J. and Fagerstrom, R. M. (1987). Marked decline in trabecular bone mineral content in healthy men with age: lack of association with sex steroid levels. *J. Am. Geriatr. Soc.*, **35**, 189–97

85. Drinka, P. J., Olson, J., Bauwens, S., Voeks, S. K., Carlson, I. and Wilson, M. (1993). Lack of association between free testosterone and bone density separate from age in elderly males. *Calcif. Tissue Int.*, **52**, 67–9

86. Wishart, J. M., Need, A. G., Horowitz, M., Morris, H. A. and Nordin, B. E. C. (1995). Effect of age on bone density and bone turnover in men. *Clin. Endocrinol.*, **42**, 141–6

87. Morley, J. E., Perry, H. M., Kaiser, F. E., Kraenzle, D., Jensen, J., Houston, K., Mattammal, M. and Perry, H. M. (1993). Effects of testosterone replacement therapy in old hypogonadal males: a preliminary study. *J. Am. Geriatr. Soc.*, **41**, 149–52

88. Tenover, J. S. (1992). Effects of testosterone supplementation in the aging male. *J. Clin. Endocrinol. Metab.*, **75**, 1092–8

89. Orwoll, E. S. and Oviatt, S. (1992). Transdermal testosterone supplementation in normal older men. Program of the 74th Annual Meeting of the Endocrine Society, San Antonio, TX (Abstract 1071) The Endocrine Society, Bethesda

DISCUSSION

Bain: We have heard from Dr Orwoll that osteoporosis in women is the focus of great attention and that we have not paid significant attention to the possibility of osteoporosis in men. How aggressive should we be in looking for osteoporosis in men, particularly in view of the discouraging results we have just heard with respect to treatment, at least in partial hypogonadism? If we find a low testosterone level, how low should it be before we treat? If we don't find low testosterone, what treatment should we offer? Having heard there is possibly no treatment, should we look for osteoporosis in the first place?

Orwoll: I think we should look for osteoporosis in men and the epidemiology that I showed you suggests that the number of fractures in men is quite significant. But to justify what we do after we find it, how to make those hard choices, is very difficult. I think it is appropriate to look for hypogonadism and if one finds frank hypogonadism – very low testosterone levels, or high gonadotropin levels – then it should obviously be treated. I don't think it justified treating age-related, mere relative hypogonadism.

There are not enough data to know whether it is useful and there are significant questions about whether it is harmful. That doesn't mean that there are no other approaches. The data that I showed you suggest that a large fraction of osteoporotic men had secondary causes for their osteoporosis (hypercortisolism, hypercalciuria, alcoholism, etc.). This leaves open a very large therapeutic window. With respect to remaining men with 'idiopathic' osteoporosis, there are no data and therefore we have to make some assumptions. Those assumptions include that we should optimize calcium and vitamin D nutrition and activity. I would also certainly take advantage of the data on anti–resorptive therapy in women under the presumption that such therapy should be effective in men as well. Some of the new bisphosphonates may offer some promise in this respect, but again, prospective data are not available.

Kaufman: I would like to comment further on this issue. The treatment of installed osteoporosis secondary to hypogonadism with androgens is somewhat disappointing because the problem in these men occurred long ago. Nevertheless, it is important to treat hypogonadal men with androgens because there is certainly reason to think that they will otherwise continue to lose more bone. Secondly I agree that we have no controlled data at all about treatments of osteoporosis in males. Therefore, we can only have a pragmatic attitude at this stage. In our practice, after having excluded secondary causes of osteoporosis such as hypercalciuria and hypogonadism, we try to find out whether we are facing a low or a high bone turnover. in the case of a low bone turnover, we will choose, extrapolating from data in women, among possibilities including fluoride, whereas in men with a high bone turnover we choose for bisphosphonates or similar substances.

Schröder: I wondered why remaining circulating testosterone in elderly men cannot prevent osteoporosis? Testosterone goes down a little bit but there are still sufficient amounts circulating. Why would that small difference cause a big difference in terms of bone parameters?

Kaufman: It is an interesting issue. We need more data before we can understand these processes. First of all, in thinking about these issues, we relied on what we believed to be the situation in women, namely that with estrogen bone mass can be maintained in women of any age. However, we do not have many data on elderly women and in fact estrogen might not be so effective in the more elderly, who may concomitantly have some degree of secondary hyperparathyroidism or marginal vitamin D deficiency. Therefore, one explanation for the apparent effects on bone of slightly decreasing androgen levels might be, as in women, that sex steroid independent changes of the elderly accentuate the effects of changes in sex steroids. On the other hand, there are some indications in postmenopausal women that even relatively small differences in adrenal androgen levels may be reflected in somewhat better bone mass. It thus remains puzzling why the androgen levels in elderly men cannot maintain bone mass.

Vermeulen: In aging men, testosterone levels decline, but estrogen levels increase. Is it possible that estrogen levels in men are an important determinant of bone mass?

Kaufman: There is certainly reason to think that some of the effects of androgens are mediated through estrogen. However, I wish to stress that the circulating estrogen levels in a male are much lower than the estrogen levels that are required in a female to maintain bone mass. Not all bone effects of androgens can be explained by the aromatization to estrogens.

Tenover: I would like to refer briefly to what I believe is one of the major issues in androgen research: is there a threshold effect for androgen therapy? This is also a relevant issue to consider in connection with the prostate, where androgen effects may become apparent possibly only above or below a certain serum threshold. In terms of bone, we have some data on correlations between bone parameters and androgen levels. However, it may be that a lot of these correlations do not work in terms of the hypogonadism of aging where changes may be more subtle. The correlational data may only be relevant to the situation above a certain threshold level. This implies that we have to look at the lower serum androgen levels, comparable to those of young hypogonadal men, and then observe what androgens are doing. The androgen threshold levels may also be dissimilar for each androgen target organ system. Therefore the picture may become very confusing; nevertheless I think that the question of androgen threshold values must unavoidably be dealt with.

Oddens: I would like to summarize some of the data presented by Dr Orwoll and then come back to that single key question. As in women, there is in men an increase in fractures with advancing age. The annual incidence of hip fractures among men aged over 65 years in the US is 4–5 per thousand. The mortality of a hip fracture patient is definitely higher in men than it is in women, *viz*. about 30%. Worldwide, an estimated 30% of all hip fractures concern men. If I make some very modest, conservative extrapolations, I would estimate that in the US about 5 to 8 million men are affected by osteoporosis, and that 2 billion US dollars are annually spent on the problem. In other words, osteoporosis in men is an enormous public health problem. Following these data, to what extent do we believe that the age-dependent decline of androgens has something to do with this enormous health problem?

Orwoll: We can clearly associate frank hypogonadism with metabolic bone disease and the risk of fracture. It is also clear that aging is associated with alterations in fracture risk as well as gonadal function. The challenge, however, is to identify the segment of the population of aging men that has osteoporosis *because* of changes in gonadal function. Only when those men can be confidently identified should we embark on testosterone replacement therapy to prevent or treat bone loss. It would be wrong to assume that all older men will benefit from androgen supplementation, but equally wrong to assume that no older men would benefit from it.

4

Androgens in relation to cardiovascular disease and insulin resistance in aging men

S. M. Haffner

GENERAL INTRODUCTION

Women *versus* men

Sex hormones appear to be major determinants of cardiovascular disease. Women have less cardiovascular disease than men, especially in the pre-menopausal period. In the general population, men have higher triglyceride and very low-density lipoprotein cholesterol (VLDL-C) concentrations than women[1]. After menopause, women have lower high-density lipo-protein cholesterol (HDL-C) and higher low-density lipoprotein cholesterol (LDL-C) than before menopause[2]. These atherogenic changes after menopause are decreased by the use of postmenopausal estrogens[2]. The differences in atherogenic risk factors between premenopausal women on the one hand and postmenopausal women and men on the other may be attributed to the presence and absence, respectively, of estrogens. Whether in men the presence of androgens plays a role as well will be reviewed in the current paper.

Effects in women

Several studies in women have shown that increased glucose and insulin concentrations are associated with increased free testosterone and decreased sex hormone binding globulin (SHBG)[3,4]. Low SHBG is thought to be an indirect measure of androgenicity, with its concentration mainly (although not exclusively) determined by the level of free estrogen and androgens[5]. Increased free testosterone and decreased SHBG in women were also associated with increased triglyceride and decreased HDL-C concentrations, which are both considered as parameters of higher cardiovascular disease risk[6–9]. Relatively weak associations of sex hormones or SHBG were observed with other parameters such as total cholesterol and LDL-C. Upper body adiposity in women has also been associated with increased free testosterone and decreased SHBG[3,10].

Women with polycystic ovary syndrome who have increased levels of free testosterone, have increased insulin and glucose levels, suggestive of insulin resistance[11,12]. Women with polycystic ovary syndrome also have increased triglyceride and decreased HDL-C levels[7]. In these women, the dyslipidemia appears to be related to both androgen excess and hyperinsulinemia[7].

Relatively few studies have been done on whether alterations in androgen levels in women predict metabolic disease. Two studies have shown that low levels of SHBG predict the incidence of non-insulin-dependent diabetes mellitus (NIDDM) in women[13,14]. These associations were not completely explained by differences in baseline insulin concentrations between converters and non-converters to diabetes.

Relatively fewer data are available on the relationship of sex hormones to metabolic variables in postmenopausal women than in premenopausal women. The associations appear to be similar, although somewhat weaker, in postmenopausal women as compared with premenopausal women. Masarei and colleagues[15] found a positive association between SHBG and HDL-C concentrations. Soler and colleagues[16] found, in univariate analysis but not in multivariate analysis, that SHBG was positively associated with HDL-C and negatively associated with triglyceride concentrations. These authors also found a significant inverse association between SHBG and insulin concentrations after adjustment for obesity and waist-to-hip ratio. Cauley and co-workers[17] did not find significant correlations of lipids and lipoproteins with either estrone or estradiol in postmenopausal women. These authors did not measure free hormone or SHBG concentrations. Significant correlations between decreased SHBG and increased triglyceride and insulin concentrations and decreased HDL-C levels, independent of obesity and body fat distribution were found by Haffner and colleagues[18]. Several studies in postmenopausal women have shown decreased SHBG to be associated with upper body adiposity[16,19]. As in premenopausal women, decreased SHBG has predicted the incidence of NIDDM in postmenopausal women[13]. In summary, the evidence available so far indicates that higher free testosterone and lower SHBG concentrations in women are associated with some parameters of higher cardiovascular disease risk, insulin resistance and NIDDM, although this evidence is more abundant for low SHBG than for increased free testosterone.

INTERVENTION DATA IN MEN

Animal studies

In animals, the effects of testosterone administration on insulin sensitivity differ in males and females. Female rats treated with testosterone have increases in insulin resistance and decreases in the proportion of red muscle fibers[20]. The situation in male rats is more complicated[21]. Castrated male rats have increased insulin resistance which is improved by low-dose (physiological replacement) testosterone administration. Treatment with high-dose testosterone, however, worsens insulin resistance.

Human studies

Anabolic steroids have been associated with increased glucose and insulin concentrations and insulin resistance[22]. Anabolic steroids also result in lower HDL-C and triglyceride concentrations and higher LDL-C[23]. However, the effect of testosterone administration to men has not led to decreased HDL-C or changes in insulin sensitivity[24–26]. There are differences in the route of administration, since anabolic steroids are usually given orally and testosterone is usually administered by intramuscular (i.m.) injection. One explanation for the different effects of oral androgens and i.m. testosterone is that the latter readily aromatizes to 17β-estradiol while oral androgens do not form potent estrogens[27]. Another explanation is related to whether administration results in more or less physiological or alternatively, very high, supraphysiological levels. Mårin and co-workers[28] have shown that testosterone (administered by the i.m. route) in middle-aged men with low androgen levels and central adiposity decreased both visceral adiposity and insulin resistance. In this report, the improvement in insulin sensitivity was greatest in subjects with the lowest testosterone levels suggesting that testosterone replacement is more likely to be beneficial in individuals with very low androgen levels than in the general population.

CROSS-SECTIONAL ASSOCIATIONS WITH RISK FACTORS IN MEN

Lipids and lipoproteins

Increased testosterone concentrations have been associated with higher HDL-C levels in most men[29-36] but not in all studies[9,37]. However, relatively few studies have considered concomitant changes in insulin sensitivity as a possible confounding variable. (An ambiguity is whether insulin is a confounder or rather part of the causal pathway and as such should not be adjusted for.) In the studies that have examined the relationship between body fat distribution and sex hormones in men, an adverse body fat distribution has not been positively associated with increased free testosterone concentrations[38-41]. In fact, in several studies, there was a weak inverse association between testosterone concentrations and upper body adiposity[38,40,41].

Several enzymes involved in HDL-C and triglyceride metabolism may be affected by sex hormones. Lipoprotein lipase is higher in women than in men while hepatic triglyceride lipase is higher in men than in women[42-44]. Estrogen therapy decreases hepatic triglyceride lipase[42,44] while anabolic steroids increase hepatic triglyceride lipase[23,45-47]. After administration of the anabolic steroid stanozolol, increases in hepatic triglyceride lipase precede the decline in HDL-C[47]. It has been suggested by Tikkanen and colleagues[48] that the influence of sex steroids on HDL-C levels is mainly mediated by changes in hepatic triglyceride lipase.

In contrast, administration of i.m. testosterone does not alter the concentration of hepatic triglyceride lipase[25,26]. Administration of testosterone which is not aromatized to 17β-estradiol leads to an increase in hepatic triglyceride lipase and decreases in HDL-C[25]. The latter observation suggests that aromatization to estradiol might be critical.

Sex hormones may also have effects on lipolysis with age, since testosterone levels decline in men and visceral fat increases[38]. Older men have decreased lipolytic activity in subcutaneous abdominal fat which is metabolically similar to intra-abdominal fat tissue[49]. Testosterone has been shown to increase lipolytic capacity[50]. Rebuffé-Scrive and co-workers[49] found that after testosterone was administered to obese men, liposynthesis at the abdominal level decreased while lipolysis increased. No change was observed in femoral adipose tissue after testosterone administration.

Recently, higher lipoprotein(a) (Lp(a)) has been associated with increased coronary heart disease (CHD)[51–55]. Lp(a) concentrations are to a large extent determined by a genetic component[56,57]. Since Lp(a) levels do not vary between men and women[58–60], it would not be expected that sex hormones have a major effect on Lp(a) concentrations. However, a number of observations suggest a possible role for sex hormones in modulating Lp(a) concentrations. Lp(a) concentrations are higher in postmenopausal women than in premenopausal women[59,61]. Postmenopausal estrogen–progestin use lowers Lp(a) levels which has been postulated as an indication that estrogen–progestin therapy may lower cardiovascular disease risk[62,63]. Pharmacological therapy with anabolic steroids in men and women[64] and estrogen therapy in men with prostatic carcinoma[65] have been shown to reduce Lp(a) concentrations. Orchiectomy in men with prostatic carcinoma has led to an increase in Lp(a)[65]. Lp(a) levels were higher in postpubertal subjects than in prepubertal subjects with insulin-dependent diabetes mellitus[66,67]. Little work has been done on the association of endogenous sex hormones and Lp(a) concentrations. In one study in men, no association was found between Lp(a) (controlling for molecular weight of apolipoprotein A and total testosterone, free testosterone, estradiol, dehydroepiandrosterone-sulfate (DHEAS) or SHBG levels[68].

Recent data indicate considerable heterogeneity in the size and density of LDL particles[69,70]. Small dense particles have been recognized as a risk factor for coronary heart disease in several studies, although this association may not be independent of triglyceride concentrations[71–73]. Small dense LDL (apolipoprotein B) particles are also associated with NIDDM, hyperinsulinemia, increased triglyceride and decreased HDL cholesterol, male gender and high carbohydrate diet[74–79]. In most populations, men have higher triglyceride level concentrations than women, which might explain part of the gender difference in LDL particle size. However, the effect of gender on LDL particle size in one study[75] was statistically independent of HDL-C and triglyceride levels, suggesting that higher triglyceride and lower HDL-C concentrations do not explain the entire gender difference in LDL particle size. Thus, sex hormones may play a role in LDL composition. Since VLDL-C is the precursor of LDL, it is possible that differing concentrations or distributions of VLDL may influence the resulting LDL particle size. Sex hormones may influence this conversion of VLDL to LDL perhaps by affecting hepatic triglyceride lipase[80] which acts in the processing of VLDL[81]. We have recently examined the relationship of endogenous sex hormones to LDL particle size in 87 normoglycemic men. Decreased SHBG, total testosterone and DHEAS levels were significantly associated with smaller dense LDL particles. Free testosterone and estradiol concentrations were not significantly related to LDL composition. After adjustment for several possible confounding variables including age, triglyceride, HDL cholesterol and insulin resistance, LDL particle size remained significantly inversely related to total testosterone and SHBG concentrations.

In summary, increased testosterone levels may, in cross-sectional studies, be associated with higher HDL-C concentrations, this effect probably being mediated by changes in hepatic triglyceride lipase. Testosterone administration may not alter HDL-C, unless non-aromatizable modalities are used (which lower HDL-C). Decreased testosterone levels are possibly associated with an adverse body fat distribution and administration in obese men resulted in increased lipolysis and decreased liposynthesis in the abdominal region. Higher Lp(a) may be associated with increased coronary heart disease. Anabolic steroids reduced Lp(a) whereas orchiectomy resulted in higher Lp(a). Nevertheless, so far no correlation has been found between normal circulating sex hormones and Lp(a) concentrations. Finally, smaller dense LDL particles, also recognized as a risk factor for coronary heart disease, were found to be related to lower total testosterone and SHBG concentrations.

Glucose and insulin concentrations and insulin resistance

In contrast to women, increased androgen concentrations in men do not appear to be associated with increased glucose and insulin concentrations (or with insulin resistance). In contrast, several studies have shown that higher total testosterone is associated with lower glucose and especially insulin concentrations. However, these associations are weak compared with the strong positive associations observed in women.

Several studies have suggested an inverse association between total testosterone and insulin concentrations[38,39,82–84]. In one study, free testosterone was also inversely associated with insulin concentrations[84]. In this report, an inverse association between insulin levels and SHBG was noted, which was greatest in individuals with the lowest levels of free testosterone suggesting a possible plateau effect. This may explain some discrepancies in the literature in that the association between testosterone or SHBG and diabetes risk (and possibly also cardiovascular risk) may be blunted in men with relatively normal androgen levels. Only a few studies have examined the association of sex hormones with direct measures of insulin resistance such as the hyperinsulinemic euglycemic clamp. Birkeland and colleagues[85] showed strong positive correlations between insulin sensitivity (the reverse of resistance) and low SHBG ($r = 0.74$, $p < 0.001$) and moderate correlations between insulin sensitivity and higher total testosterone ($r = 0.43$, $p < 0.05$) and negative correlations with waist-to-hip ratio (WHR; $r = -0.42$, $p < 0.01$) in 23 men with NIDDM. These authors did not show a significant association between insulin sensitivity and free testosterone ($r = -0.06$), free estradiol ($r = -0.03$) or body fat distribution (BMI; $r = -0.39$). The authors also showed that the association between SHBG and insulin sensitivity remained statistically significant after adjustment for obesity and body fat distribution. Peiris and co-workers[86] showed a significant association between low SHBG and insulin secretory pulse ($r = 0.86$, $p < 0.05$) but not with peripheral insulin sensitivity in 10 non-diabetic men. Recently, Haffner and colleagues[87] have used indirect calorimetry combined with the euglycemic clamp to assess oxidative and non-oxidative glucose metabolism in relation to sex hormones in 87 normoglycemic men. They showed a statistically significant association between total and non-oxidative glucose disposal (indicating lower resistance) with SHBG ($r = 0.37$ and $r = 0.32$, respectively, both $p < 0.001$). In contrast, glucose oxidation was not significantly related to any of the sex hormones or dehydroepiandrosterone sulfate. Although defects in both glucose oxidation and non-oxidative glucose disposal are present in subjects with established NIDDM[88], non-diabetic relatives of subjects with NIDDM have decreased non-oxidative glucose disposal but normal glucose oxidation[89,90]. The data of Haffner and colleagues[87] suggest that perhaps decreased total testosterone and SHBG and increased upper body adiposity might comprise an early lesion associated with reduced non-oxidative glucose disposal in the prediabetic state. Nestler[91] has proposed that low SHBG may be a marker of insulin resistance. In summarizing these various reports, it might be postulated that low SHBG and low testosterone may be associated with higher insulin resistance in men.

Sex hormones and prevalence of NIDDM in men

Total testosterone and SHBG concentrations are significantly lower in men with NIDDM than in normoglycemic men even after matching for obesity[92–94]. Interestingly, in the latter report, testosterone levels in women with NIDDM were higher than in normoglycemic women.

PROSPECTIVE STUDIES OF NIDDM IN MEN

While low SHBG was not a significant predictor of the development of NIDDM in a small study of men[14] both low SHBG and testosterone predicted the development of NIDDM in a much larger number of men in the Multiple Risk Factor Intervention Trial[95].

PROSPECTIVE STUDIES OF CARDIOVASCULAR DISEASE IN MEN

A large number of studies have examined the cross-sectional association of sex hormones with cardiovascular disease in men. These studies have yielded mixed results. In this review, we will concentrate on prospective studies, which have a stronger research design. Barrett-Connor and Khaw[96] studied 1009 Caucasian men (aged 40–59) for 12 years. One hundred and fourteen subjects died of cardiovascular disease. Cardiovascular mortality was not significantly related to testosterone, androsterone, estrone, or estradiol.

Cauley and colleagues[97] examined in a case–control study nested within the Multiple Risk Factor Intervention Trial the relationship between sex hormones and subsequent risk of a major coronary event in 163 men who had such an event and 163 controls. Cases and controls were matched for age, serum cholesterol level, randomization group, randomization date and clinic. The follow-up period was 6–8 years. There were no significant differences between cases and controls for total and free testosterone, total and free estradiol, androsterone and estrone concentrations.

We recently examined the relation of total and free testosterone, SHBG, DHEAS, estrone and estradiol in a nested case–control study from the Wisconsin Epidemiologic Study of Diabetic Retinopathy (WESDR) in which 52 male subjects who died of ischemic heart disease were matched to 104 male subjects who remained free of ischemic heart disease[98]. The follow-up period was 4 years. None of the sex hormones were significantly related to ischemic heart disease.

TEMPORAL ASSOCIATIONS

One of the problems in the area of cardiovascular risk factors and sex hormones has been that it is difficult in cross-sectional studies to determine the direction of causality. This applies in particular to associations found between sex hormones and insulin resistance. In fact, both types of causal association have been postulated, *vis* sex hormones influence insulin (resistance) and insulin may itself determine sex hormone concentrations. Two examples follow: anabolic steroid use is associated with increased insulin resistance[22]; while Barbieri and Ryan[99] have reported that insulin infusions increase androgen output by the ovary. Recently, Pasquali and co-workers[100] have shown that insulin can stimulate testosterone production and simultaneously inhibit SHBG concentrations in both normal weight and obese men. *In vitro*, insulin has also been shown to suppress the production of SHBG in human hepatoma cell line[101].

CONCLUSIONS

We have shown in this review that decreased testosterone concentrations are associated with certain cardiovascular risk factors (especially insulin resistance, decreased HDL-C and increased triglyceride levels and smaller denser LDL-C) in cross-sectional studies. Orchiectomy was associated with higher Lp(a) concentrations, but low testosterone within the physiological range was not. Neither was low testosterone strongly associated with blood pressure or the absolute concentration of LDL-C. In prospective studies, low testosterone and SHBG were significantly related to incidence of NIDDM in men but not to the incidence of coronary heart disease. Short-term intervention studies have shown an improvement with testosterone administration on insulin sensitivity and dyslipidemia in both men with low levels of circulating androgens and in animal studies.

A number of paradoxes may be suggested.

(1) Low testosterone is associated with cardiovascular risk factors but not with cardiovascular disease. (This is similar to insulin resistance which is strongly associated with cardiovascular risk factors but not with coronary heart disease.)

(2) Are there protective factors (not easily identified) of having low testosterone levels? (A parallel here is that insulin resistance in a few studies is associated with lower Lp(a) levels.)

(3) Is there a threshold effect below which the testosterone associations are stronger, for example in men with very low testosterone concentrations as has been suggested by work by Mårin and co-workers[28] and Haffner and co-workers[84]?

In conclusion, testosterone replacement in men with below normal testosterone is likely to decrease the risk of NIDDM. However, whether this also reduces CHD is unknown, (at least, resulting testosterone concentrations within the physiological range may not increase CHD risk). In the general male population, including men with normal testosterone levels, testosterone replacement is unlikely to have an effect on CHD. Clinical intervention studies, which are much stronger as a research design than are cross-sectional or prospective observational studies, are warranted to investigate these issues further.

REFERENCES

1. Heiss, G., Tamir, I., Davis, C. E., Tyroler, H. A., Rifkand, B. M., Schonfeld, G., Jacobs, D. and Frantz, I. D. Jr. (1980). Lipoprotein cholesterol distributions in selected North American populations: the Lipid Research Clinics Program Prevalence Study. *Circulation*, **61**, 302–15

2. Matthews, K. A., Meilahn, E., Kuller, L. H., Kelsey, S. F., Caggiula, A. W. and Ring, R. R. (1989). Menopause and risk factors for coronary heart disease. *N. Engl. J. Med.*, **321**, 641–6

3. Evans, D. J., Hoffman, R. G., Kalkhoff, R. K. and Kissebah, A. H. (1983). Relationship of androgenic activity to body fat topography, fat cell morphology and metabolic aberrations in premenopausal women. *J. Clin. Endocrinol. Metab.*, **57**, 304–10

4. Haffner, S. M., Katz, M. S., Stern, M. P. and Dunn, J. F. (1988). The relationship of sex hormones to hyperinsulinemia and hyperglycemia. *Metabolism*, **37**, 683–8

5. Anderson, D. C. (1974). Sex hormone binding globulin. *Clin. Endocrinol.*, **3**, 69–96

6. Haffner, S. M., Katz, M. S., Stern, M. P. and Dunn, J. R. (1989). Association of decreased sex hormone binding globulin and cardiovascular risk factors. *Arteriosclerosis*, **9**, 136–43

7. Wild, R. A., Applebaum-Bowden, D., Demers, L. M., Bartholomew, M., Landis, J. R., Hazzard, W. R. and Santen, R. J. (1990). Lipoprotein, lipids in women with androgen excess: independent associations with increased insulin and androgen. *Clin. Chem.*, **36**, 283–9

8. Gorbach, S. L., Schaefer, E. J., Woods, M., Loncope, C., Dwyer, J. T., Goldin, B. R., Morrill-LaBrode, A. and Dallal, G. (1989). Plasma lipoprotein cholesterol and endogenous sex hormones in healthy young women. *Metabolism*, **38**, 1077–81

9. Semmens, J., Rouse, J., Beilin, L. J. and Masarei, J. R. (1983). Relationship of plasma HDL cholesterol to testosterone, estradiol and sex hormone binding globulin levels in men and women. *Metabolism*, **32**, 428–32

10. Haffner, S. M., Katz, M. S., Stern, M. P. and Dunn, J. F. (1989). Relationship of sex hormone binding globulin to overall adiposity and body fat distribution in a biethnic population. *Int. J. Obesity*, **13**, 1–9

11. Burghen, G. A., Givens, J. R. and Kitabchi, A. E. (1980). Correlation of hyperandrogenism with hyperinsulinism in polycystic ovarian disease. *J. Clin. Endocrinol. Metab.*, **50**, 113–16

12. Smith, S., Ravnikar, V. A. and Barbieri, R. L. (1987). Androgen and insulin response to an oral glucose challenge in hyperandrogenic women. *Fertil. Steril.*, **48**, 72–7

13. Lindstedt, G., Lundberg, P. A., Lapidus, L., Lundgren, H., Bengtsson, C. and Björntorp, P. (1980). Low sex hormone binding globulin concentrations as independent risk factor for development of NIDDM: 12-yr. follow-up of population study of women in Gothenburg, Sweden. *Diabetes*, **40**, 123–8

14. Haffner, S. M., Valdez, R. A., Morales, P. A., Hazuda, H. P. and Stern, M. P. (1993). Decreased sex hormone binding globulin predicts non-insulin dependent diabetes mellitus in women but not in men. *J. Clin. Endocrinol. Metab.*, **77**, 56–60

15. Masarei, J. R. L., Armstrong, B. K., Skinner, M. W., Ratajczak, T., Hähnel, R., Crooke, D. and Clarke, H. T. (1980). HDL cholesterol and sex hormone status. (Letter). *Lancet*, **1**, 208

16. Soler, J. T., Folsom, A. R., Kaye, S. A. and Prineas, R. J. (1989). Associations of abdominal adiposity, fasting insulin, sex hormone binding globulin, and estrone with lipids and lipoproteins in postmenopausal women. *Atherosclerosis*, **79**, 21–7

17. Cauley, J. A., Gatai, J. P., Kuller, L. H. and Powell, J. G. (1990). The relationship of endogenous sex steroid hormone concentrations to serum lipid and lipoprotein levels in postmenopausal women. *Am. J. Epidemiol.*, **132**, 884–94

18. Haffner, S. M., Katz, M. S. and Dunn, J. F. (1992). Relationship of sex hormone binding globulin to lipid, lipoprotein, glucose and insulin concentrations in postmenopausal women. *Metabolism*, **42**, 278–84

19. Haffner, S. M., Katz, M. S. and Dunn, J. F. (1991). Increased upper body and overall adiposity is associated with decreased sex hormone binding globulin in postmenopausal women. *Int. J. Obesity*, **15**, 471–8

20. Holmäng, A., Svedberg, J., Jennische, E. and Björntorp, P. (1990). Effects of testosterone on muscle, insulin sensitivity and morphology in female rats. *Am. J. Physiol.*, **259**, E555–60

21. Holmäng, A. and Björntorp, P. (1992). The effects of testosterone on insulin sensitivity in male rats. *Acta Physiol. Scand.*, **146**, 505–10

22. Shoupe, D. and Lobo, R. A. (1984). The influence of androgens on insulin resistance. *Fert. Steril.*, **41**, 385–8

23. Taggert, H., Applebaum-Bowden, D., Haffner, S. M., Warnick, G. R., Cheung, M. C., Albers, J. J., Chestnut, C. H. (ed). and Hazzard, W. R. (1984). Reduction in high density lipoproteins by anabolic steroids (stanozolol) therapy for postmenopausal osteoporosis. *Metabolism*, **31**, 1147–52

24. Friedel, K. E., Jones, R. E., Hannan, C. J. Jr. and Plymate, S. R. (1989). The administration of pharmacological doses of testosterone or 19-nortestosterone to normal men is not associated with increased insulin secretion or impaired glucose tolerance. *J. Clin. Endocrinol. Metab.*, **68**, 971–5

25. Friedel, K. E., Hannan, C. J. Jr., Jones, R. E. and Plymate, S. R. (1990). High density lipoprotein cholesterol is not decreased if aromatizable androgen is administered. *Metabolism*, **39**, 69–74

26. Thompson, P. D., Cuillinane, E. M., Sady, S. P., Chenevert, C., Saritelli, A. L. and Hebert, P. N. (1989). Contrasting effects of testosterone and stanozolol on serum lipoprotein levels. *J. Am. Med. Assoc.*, **261**, 1165–8

27. Dimick, D. F., Heron, M., Baulieu, E. E. and Jayle, M. F. (1961). A comparative study of the metabolic fate of testosterone, 17 α-methyl testosterone, 19-nortestosterone, 17 α-methyl-19-nortestosterone and 17-α-methyl-estr-5(10)-ene=17-β-01-3 one in normal males. *Clin. Chem. Acta*, **6**, 63–71

28. Mårin, P., Holmäng, S., Jönsson, L., Sjostrom, L., Kvist, H., Holm, G., Lindstedt, G. and Björntorp, P. (1992). The effects of testosterone treatment

on body composition and metabolism in middle-aged obese men. *Int. J. Obesity*, **16**, 991–7

29. Gutai, J., LaPorte, R., Kuller, L., Dai, W., Falvo-Gerard, L. and Caggiula, A. (1981). Plasma testosterone, high density lipoprotein cholesterol and other lipoprotein fractions. *Am. J. Cardiol.*, **48**, 897–902

30. Heller, R. F., Wheeler, M. J., Micallef, J., Miller, N. E. and Lewis, B. (1983). Relationship of high density lipoprotein cholesterol with total and free testosterone and sex hormone binding globulin. *Acta Endocrinol. (Copenh.)*, **104**, 253–6

31. Dai, W. S., Gutai, J. P., Kuller, L. H., LaPorte, R. E., Falvo-Gerard, L. and Caggiula, A. (1984). Relation between plasma high density lipoprotein cholesterol and sex hormone concentrations in men. *Am. J. Cardiol.*, **53**, 1259–63

32. Hamäläinen, E., Aldercreutz, H., Enholm, C. and Puska, P. (1986). Relationships of serum lipoproteins and apolipoproteins to sex hormones and the binding capacity of sex hormone binding globulin in healthy Finnish men. *Metabolism*, **35**, 535–41

33. Duell, P. B. and Bierman, E. I. (1990). The relationship between sex hormones and high density lipoprotein cholesterol levels in healthy adult men. *Arch. Intern. Med.*, **150**, 2317–20

34. Khaw, K. T. and Barrett-Connor, E. (1991). Endogenous sex hormones, high density lipoprotein cholesterol, and other lipoprotein fractions in men. *Arterioscler. Thromb.*, **11**, 489–94

35. Lichtenstein, M. J., Yarnell, J. W., Elwood, P. C., Beswick, A. D., Sweetnam, P. M., Marks, V., Teale, D. and Riad-Fahmy, D. (1987). Sex hormones, insulin, lipids, and prevalent ischemic heart disease. *Am. J. Epidemiol.*, **126**, 647–57

36. Haffner, S. M., Mykkänen, L., Valdez, R. A., Stern, M. P. and Katz, M. S. (1993). Relationship of sex hormones to lipids and lipoproteins in non-diabetic men. *J. Clin. Endocrinol. Metab.*, **77**, 1610–15

37. Stefanick, M. L., Williams, P. T., Krauss, R. M., Terry, R. B., Vranizan, K. M. and Wood, P. D. (1987). Relationships of plasma estradiol, testosterone and sex hormone binding globulin with lipoproteins, apolipoproteins, and high density lipoprotein subfractions in men. *J. Clin. Endocrinol.*, **64**, 723–9

38. Seidell, J. C., Björntorp, P., Kvist, H., Sannerstedt, R. and Sjostrom, L. (1990). Visceral fat accumulation in men is positively associated with insulin,

glucose, and C-peptide levels, but negatively with testosterone levels. *Metabolism*, **39**, 897–901

39. Pasquali, R., Casimirri, F., Cantobelli, S., Melchionda, N., Morselli Labate, A. M., Fabbri, R., Capelli, M. and Bortoluzzi, L. (1991). Effect of obesity and body fat distribution on sex hormones and insulin in men. *Metabolism*, **40**, 101–4

40. Khaw, K. T. and Barrett-Connor, E. (1992). Lower endogenous androgens predict central adiposity in men. *Ann. Epidemiol.*, **2**, 675–82

41. Haffner, S. M., Valdez, R. A., Stern, M. P. and Katz, M. S. (1993). Obesity, body fat distribution and sex hormones in men. *Int. J. Obesity*, **17**, 643–9

42. Krauss, R. M., Levy, R. I. and Fredrickson, D. S. (1974). Selective measurement of two lipase activities in postheparin plasma from normal subjects and patients with hyperlipoproteinemia. *J. Clin. Invest.*, **54**, 1107–24

43. Applebaum-Bowden, D., Haffner, S. M., Wahl, P. W., Hoover, J. J., Hornick, G. R., Albert, J. J. and Hazzard, W. R. (1985). Post heparin plasma triglyceride lipases. Relationships with low density lipoprotein triglyceride and high density lipoprotein cholesterol. *Arteriosclerosis*, **5**, 273–8

44. Applebaum, D. M., Goldberg, A. P., Pykalisto, O. J., Brunzell, J. D. and Hazzard, W. R. (1977). Effect of estrogen on post-heparin lipolytic activity. Selective decline in hepatic triglyceride lipase. *J. Clin. Invest.*, **59**, 601–8

45. Enholm, C., Huttunen, J. K., Kinnunen, P. J., Miettinen, T. A. and Nikkila, E. A. (1975). Effect of oxandralone treatment on the activity of lipoprotein lipase, hepatic lipase and phospholipase A of human postheparin plasma. *N. Engl. J. Med.*, **292**, 1314–17

46. Haffner, S. M., Kushwaha, R. S., Foster, D. M., Applebaum-Bowden, D. and Hazzard, W. R. (1983). Studies on the metabolic mechanism of reduced high density lipoproteins during anabolic steroid therapy. *Metabolism*, **32**, 413–20

47. Applebaum-Bowden, D. M., Haffner, S. M. and Hazzard, W R. (1987). The dyslipoproteinemia of anabolic steroid therapy increase in hepatic triglyceride lipase precedes the decrease in high density lipoprotein cholesterol. *Metabolism*, **36**, 949–52

48. Tikkanen, M. J., Nikkila, E. A., Kuusi, T. and Sipinen, S. (1982). High density lipoprotein-2 and hepatic lipase: reciprocal changes produced by estrogen and norgesterol. *J. Clin. Endocrinol. Metab.*, **54**, 1113–17

49. Rebuffé-Scrive, M., Lonnroth, P., Mårin, P., Wesslau, C., Björntorp, P. and Smith, U. (1987). Regional adipose tissue metabolism in men and postmenopausal women. *Int. J. Obesity*, **11**, 347–55

50. Xu, X., DePergola, G. and Björntorp, P. (1990). The effects of androgens on the regulation of lipolysis in adipose precursor cells. *Endocrinology*, **126**, 1229–34

51. Berg, K., Dahlen, G. and Frick, M. H. (1975). Lp(a) lipoprotein and prebeta lipoprotein in patients with coronary heart disease. *Clin. Genet.*, **8**, 183–9

52. Rhoads, G. G., Dahlen, G., Berg, K., Morton, N. E. and Dannenberg, A. L. (1986). Lipoprotein (a) as a risk factor for myocardial infarction. *J. Am. Med. Assoc.*, **256**, 2540–4

53. Darrington, P. N., Ishola, M., Hunt, L., Arrol, S. and Bhatnager, D. (1988). Apolipoproteins(a), AI, and B, and parental history in men with early onset ischemic heart disease. *Lancet*, **1**, 1070–3

54. Dahlen, G., Guyton, J. R., Attar, M., Farmer, J. A., Kautz, J. A. and Gotto, A. M. Jr. (1986). Association of levels of lipoproteins Lp(a), plasma lipids, and other lipoproteins with coronary artery disease documented by angiography. *Circulation*, **74**, 758–65

55. Seed, M., Hoppichler, F., Reavley, D., McCarthy, S., Thompson, G. R., Boerwinkle, E. and Utterman, G. (1990). Relationship of serum lipoprotein(a) concentration and apolipoprotein(a) phenotype to coronary heart disease in patients with familial hypercholesterolemia. *N. Engl. J. Med.*, **322**, 1494–9

56. Utermann, G., Menzel, H. J., Kraft, H. G., Duba, H. C., Kemmler, H. G. and Seitz, C. (1987). Lp(a) glycoprotein phenotypes. Inheritance and relation to Lp(a) lipoprotein concentrations in plasma. *J. Clin. Invest.*, **80**, 458–65

57. Rainwater, D. L., Manis, G. S. and Vandeberg, J. L. (1989). Hereditary and dietary effects on apolipoproteins(a) isoforms and Lp(a) in baboons. *J. Lipid Res.*, **30**, 549–58

58. Haffner, S. M., Gruber, K. K., Morales, P. A., Hazuda, H. P., Valdez, R. A., Mitchell, B. D. and Stern, M. P. (1992). Lipoprotein(a) concentrations in Mexican Americans and non-Hispanic whites: the San Antonio Heart Study. *Am. J. Epidemiol.*, **136**, 1060–8

59. Guyton, J. R., Dahlen, G. H., Patsch, W., Kautz, J. H. and Gotto, A. M. Jr. (1985). Relationship of plasma lipoprotein Lp(a) levels to race and to apolipoprotein B. *Arteriosclerosis*, **5**, 265–72

60. Schriewer, H., Assman, G., Sandkamp, M. and Schulte, H. (1984). The relationship of lipoprotein(a) (Lp(a)) to risk factors for coronary heart disease: initial results of the prospective epidemiology study on company employees in Westfalia. *J. Clin. Chem.*, **22**, 591–6

61. Sandkamp, M. and Assman, G. (1990). Lipoprotein(a) in PROCAM participants and young myocardial infarction survivors. In Scanu, A. M. (ed.) *Lipoprotein(a): 25 years of Progress*, pp. 205–10. (Orlando: Academic Press)

62. Soma, M., Fumagalli, R., Paoletti, R., Meschia, M., Miani, M. C., Crosignani, P., Ghanem, K., Guabatz, J. and Morrisett, J. D. (1991). Plasma Lp(a) concentration after estrogen and progesterone in postmenopausal women (Letter). *Lancet*, **337**, 612

63. Soma, M. R., Osnago-Gadda, I., Paoletti, R., Fumagalli, R., Morrisett, J. D., Meschia, M. and Crosignani, P. (1993). The lowering of lipoprotein(a) induced by estrogen plus progesterone replacement in postmenopausal women. *Arch. Intern. Med.*, **153**, 1462–8

64. Albers, J. J., Taggert, H. M., Applebaum-Bowden, D., Haffner, S. M., Chestnut, C. H. and Hazzard, W. R. (1984). Reduction of lecithin cholesterol acyltransferase, apolipoprotein D and the Lp(a) lipoprotein with the anabolic steroid stanozolol. *Biochem. Biophys. Acta*, **795**, 293–6

65. Henriksson, P., Angelin, B. and Berglund, L. (1992). Hormonal regulation of serum Lp(a) levels: opposite effects after estrogen treatment and orchidectomy in males with prostatic carcinoma. *J. Clin. Invest.*, **89**, 1166–71

66. Couper, J. J., Bates, D. J., Cocciolone, R., Magarey, A. M., Boulton, T. J., Penfold, J. L. and Ryall, R. G. (1993). Association of lipoprotein(a) with puberty in insulin dependent diabetes. *Diabetes Care*, **16**, 869–73

67. Haffner, S. M., Frangos, M., Williamson, J., Santiago, J., Valdez, R., Aldrete, G., Mykkänen, L., Gruber, K. K. and Rainwater, D. L. (1994). Lp(a) concentrations in children with insulin dependent diabetes mellitus. *Chem. Phys. Lipids*, **67/68**, 223–31

68. Haffner, S. M., Mykkänen, L., Gruber, K. K., Rainwater, D. L. and Laakso, M. (1994). Lack of association between sex hormones and Lp(a) concentrations in American and Finnish men. *Arterioscler. Thromb.*, **14**, 19–24

69. Shen, M. M., Krauss, R. M., Lindgren, F. T. and Forte, T. M. (1981). Heterogeneity of serum low density lipoproteins in normal human subjects. *J. Lipid Res.*, **22**, 236–44

70. Krauss, R. M. and Burke, D. J. (1982). Identification of multiple subclasses of plasma low density lipoproteins in normal humans. *J. Lipid Res.*, **23**, 97–104

71. Crouse, J. R., Parks, J. S., Schey, H. M. and Kahl, F. R. (1985). Studies of low density lipoprotein molecular weight in human beings with coronary artery disease. *J. Lipid Res.*, **26**, 566–74

72. Austin, M. A., Breslow, J. L., Hennekens, C. H., Buring, J. H., Willett, W. C. and Krauss, R. M. (1988). Low density lipoprotein subclass patterns and risk of myocardial infarction. *J. Am. Med. Assoc.*, **260**, 1917–21

73. Campos, H., Genest, J. J., Blijlevens, E., McNamara, J. R., Jenner, J. L., Ordovas, J. M., Wilson, P. W. and Schaefer, E. J. (1992). Low density lipoprotein particle size and coronary artery disease. *Arteriosclerosis*, **12**, 187–95

74. Barakat, H. A., Carpenter, J. W., McLendon, V. D., Khazanie, P., Leggett, N., Heath, J. and Marks, R. (1990). Influence of obesity impaired glucose tolerance, and NIDDM on LDL structure and composition: possible link between hyperinsulinemia and atherosclerosis. *Diabetes*, **39**, 1527–33

75. McNamara, J. R., Campos, H., Ordovas, J. M., Peterson, R. M., Wilson, P. W. F. and Schaefer, E. J. (1987). Effect of gender, age and lipid status on low density subfraction distribution: results of the Framingham Offspring Study. *Arteriosclerosis*, **7**, 483–90

76. Swinkels, D. W., Demacker, P. N., Hendriks, J. C. and Van't Laur, A. (1989). Low density lipoprotein subfractions and relationships of other risk factors for coronary artery disease in healthy individuals. *Arteriosclerosis*, **9**, 604–13

77. Selby, J. V., Austin, M. A., Newman, B., Ahang, D., Quesenberry, C. P. Jr, Mayer, E. J. and Krause, R. M. (1993). LDL subclass phenotypes and insulin resistance syndrome in women. *Circulation*, **88**, 381–7

78. Campos, H., Willett, W. C., Peterson, R. M., Siles, X., Bailey, S. M., Wilson, P. W., Posner, B. M., Ordovas, J. M. and Schaefer, E. J. (1991). Nutrient intake comparisons between Framingham and rural and urban Puriscal, Costa Rica: associations with lipoproteins, apolipoproteins and low density lipoprotein particle size. *Arterioscler. Thromb.*, **11**, 1089–99

79. Haffner, S. M., Mykkänen, L., Valdez, R. A., Paidi, M., Stern, M. P. and Howard, B. V. (1993). LDL size and subclass pattern in a biethnic population. *Arteriosclerosis Thromb.*, **13**, 1623–30

80. Kinnunen, P. H. H., Vitanen, J. A. and Vianio, P. (1983). Lipoprotein lipase and hepatic endothelial lipase: their roles in plasma lipoprotein metabolism. *Atherosclerotic Rev.*, **11**, 65–105

81. Goldberg, I. J., Le, N. A., Paterniti, J. R., Ginsberg, H. N., Lindgren, F. T. and Brown, W. V. (1982). Lipoprotein metabolism during acute inhibition of hepatic triglyceride lipase in the cynomolgus monkey. *J. Clin. Invest.*, **70**, 1184–92

82. Simon, D., Preziosi, P., Barrett-Connor, E., Roger, M., Saint-Paul, M., Nahoul, K. and Papoz, L. (1992). Interrelationship between plasma testosterone and plasma insulin in healthy adult men: the Telecom study. *Diabetologia*, **35**, 173–7

83. Phillips, G. (1993). Relationship between sex hormones and the glucose insulin lipid defect in men with obesity. *Metabolism*, **42**, 116–20

84. Haffner, S. M., Valdez, R. A., Mykkänen, L., Stern, M. P. and Katz, M. S. (1994). Decreased testosterone and dehydroepiandrosterone sulfate are associated with increased glucose and insulin concentrations in non-diabetic men. *Metabolism*, **43**, 599–603

85. Birkeland, K. I., Hanssen, K. F., Torjesen, P. A. and Stein Vaaler, S. (1993). Low level of sex-hormone binding globulin is positively correlated with insulin insensitivity in men with type II diabetes. *J. Clin. Endocrinol. Metab.*, **76**, 275–8

86. Peiris, A. N., Stagner, J. I. Plymate, S. R., Vogel, R. L., Heck, M. and Samols, E. (1993). Relationships of insulin secretory pulses to sex hormone binding globulin in normal men. *J. Clin. Endocrinol. Metab.*, **76**, 279–82

87. Haffner, S. M., Karhapää, P., Mykkänen, L. and Laakso, M. (1994). Insulin resistance, body fat distribution and sex hormones in men. *Diabetes*, **43**, 212–19

88. DeFronzo, R. A. (1988). Lilly Lecture 1987. The triumvirate: beta cell, muscle, liver: a collusion responsible for NIDDM. *Diabetes*, **37**, 667–87

89. Eriksson, J., Fransilla-Kullunki, A., Ekstrand, A., Saloranta, C., Widen, E., Schalin, C. and Groop, L. (1989). Early metabolism defects in persons at increased risk for NIDDM. *N. Engl. J. Med.*, **321**, 337–43

90. Gulli, G., Ferrannini, E., Stern, M., Haffner, S. M. and DeFronzo, R. A. (1992). The metabolic profile of NIDDM is fully established in glucose intolerant offspring of two Mexican American parents. *Diabetes*, **41**, 1575–86

91. Nestler, J. E. (1993). Sex hormone binding globulin: a marker for hyperinsulinemia and/or insulin resistance (Editorial). *J. Clin. Endocrinol. Metab.*, **76**, 273–4

92. Barrett-Connor, E., Khaw, K. T. and Yen, S. S. (1990). Endogenous sex hormone levels in older adult men with diabetes mellitus. *Am. J. Epidemiol.*, **132**, 895–901

93. Barrett-Connor, E. (1992). Lower endogenous androgen levels and dyslipidemia in men with non-insulin dependent diabetes mellitus. *Ann. Intern. Med.*, **117**, 807–11

94. Andersson, B., Vermeulen, A., Mårin, P., Björntorp, P. and Lissner, L. (1994). Testosterone concentrations in women and men with NIDDM. *Diabetes Care*, **17**, 405–11

95. Haffner, S. M., Shaten, J. S., Stern, M. P., Smith, G. D. and Kuller, L. (1996). Low levels of sex hormone binding globulin and testosterone predict the development of non-insulin dependent diabetes mellitus in men. *Am. J. Epidemiol.*, **143**, 889–97

96. Barrett-Connor, E. and Khaw, K. S. (1988). Endogenous sex hormones and cardiovascular disease in men. A prospective population-based study. *Circulation*, **78**, 539–45

97. Cauley, J. A., Gutai, J. P., Kuller, L. H. and Dai, W. S. (1987). Usefulness of sex steroid hormone levels in predicting coronary artery disease in men. *Am. J. Cardiol.*, **60**, 771–7

98. Haffner, S. M., Moss, S. E., Klein, B. E. K. and Klein, R. (1996). Sex hormones and DHEA-SO4 in relation to ischemic heart disease in diabetic subjects: the WESDR study. *Diabetes Care*, (in press)

99. Barbieri, R. L. and Ryan, K. S. (1983). Hyperandrogenism, insulin resistance and acanthosis nigricans syndrome: a common endocrinopathy with direct pathophysiologic features. *Am. J. Obstet. Gynecol.*, **147**, 90–101

100. Pasquali, R., Casimirri, F., De Lasio, R., Meisini, P., Boshi, S., Chierici, R., Flamia, R., Biscotti, M. and Viaennati, V. (1995). Insulin regulates testosterone and sex hormone-binding globulin concentrations in adult normal weight men and obese men. *J. Clin. Endocrinol. Metab.*, **80**, 654–8

101. Plymate, S. R., Matej, L. A., Jones, R. E. and Friedel, K. E. (1988). Inhibition of sex hormone binding globulin production in a human hepatoma (hep G2) cell line by insulin and prolactin. *J. Clin. Endocrinol. Metab.*, **67**, 460–4

5

Sex hormones, aging, ethnicity and insulin sensitivity in men: an overview of the TELECOM Study

D. Simon, K. Nahoul and M.-A. Charles

The TELECOM Study was designed in the early 1980s to assess the ability of the marker glycated hemaglobin A1c (HbA1c) to be used as a tool for diabetes diagnosis and to investigate the relationship between sex hormones and cardiovascular risk. It was planned to perform a prospective survey in order to compare HbA1c and plasma glucose values as predictors of specific diabetic complications and to evaluate sex hormones as predictors of cardiovascular diseases. The TELECOM Study recruited healthy adult subjects of both sexes. We report in this paper the data which have been collected concerning sex hormones, aging, ethnicity and insulin sensitivity in men. Some results have already been published[1-3].

SUBJECTS AND METHODS

From April 1985 to July 1987, 1718 healthy, adult, non–diabetic men, who voluntarily attended a center for preventive medicine, were examined in a cross–sectional study. The participants were France Telecom employees working in the Paris area, who are offered a free medical checkup every 5 years for those less than 40 years of age and every 2 years for those over 40. Most were sedentary office workers. Among them, 186 were black, mainly from the French Caribbean islands (referred to here as Afro–Caribbean), and 1532 were white (referred to here as Caucasian).

A self-administered questionnaire, filled out at home and reviewed by a secretary with the subject during the consultation, inquired about history of chronic disease, tobacco and alcohol consumption. A physical examination recorded height, weight and subscapular skinfold thickness. Blood from fasting subjects was obtained between 08.00 and 09.00 on the day of the consultation. Sex hormones were measured by radioimmunoassay after column chromatography, on celite for plasma total testosterone (PTT)[4] and on Sephadex LH 20 for plasma estrone and estradiol[5]. Plasma insulin was measured by radioimmunoassay with the INSI-PR method (CIS Bio-Industrie, Gif-sur-Yvette, France). The samples were frozen at $-20\,°C$ until hormones were assayed an average of 32 days later. Plasma glucose was measured fasting and 2 h after a 75-g oral glucose load.

In order to investigate further the relationship between PTT and cardiovascular risk factors, a nested case–control study was performed in 1992–3. Two groups of 30 subjects without antidiabetic or hypolipidemic drugs, contrasted by their PTT level, and matched on age and ethnic group, were constituted. They were asked to come at 08.30, fasting from midnight. They underwent a new clinical examination by a physician who measured blood pressure in supine position, waist and hip circumferences. Fasting blood was drawn between 08.30 and 09.30 then a standard 75-g oral glucose tolerance test was performed. In addition to parameters measured in the previous cross-sectional study, high–density lipoprotein (HDL)-cholesterol, apoA1 and apoB lipoproteins, and 2-h plasma insulin were also measured. Moreover, plasma bioavailable testosterone and sex hormone-binding globulin (SHBG) were determined according to a modification of the technique of Tremblay and Dube[6] and by electroimmunodiffusion using the kit SBP-Film (Sebia, France) respectively.

Furthermore, in 1994–5, among the subjects with low PTT re-examined in 1992–3, 18 agreed to participate in a controlled randomized clinical trial to compare testosterone (T), dihydrotestosterone (DHT) and placebo (P) gels during a 3-month period, with metabolic parameters (fasting and 2-h plasma glucose and insulin, fasting lipids) as main end-points. These successive protocols had been approved by the Ethics Committee and all the participants gave informed written consent.

Statistical analysis was conducted with the help of the SAS and SPSS softwares to perform mean comparisons by variance analysis, with adjustment for significant covariates using analysis of covariance, and to calculate Pearson correlation coefficients. For variables with distributions positively skewed, calculations were performed after logarithmic transformation. Concerning the clinical trial, non-parametric tests were used given the small sample size. A 2-tailed p value of ≤ 0.05 was used to indicate statistical significance. Mean and standard error of the mean [m(SEM)], or mean followed by the 95% confidence interval in brackets when logarithmic transformation was needed, are given.

RESULTS

Table 1 describes the two ethnic groups in this male population, showing a significantly higher level of fasting plasma insulin in Afro-Caribbeans[3]. PTT was not significantly different: 6.23 (SEM 0.15) vs. 6.11 (0.05) ng/ml in Afro-Caribbeans and Caucasians respectively ($p = 0.42$) while plasma estrone and estradiol were significantly higher in Afro-Caribbeans, even after adjustment for significant covariates ($p = 0.0001$ and $p < 0.0001$ respectively)[7]. Table 2 shows the matrix of correlations between sex hormones and age, body mass index, systolic blood pressure, metabolic parameters, fasting plasma insulin and γ-glutamyl-transferase (γGT). It indicates significant negative relationships for PTT with age, body mass index, systolic blood pressure, all the metabolic cardiovascular risk factors and γGT. On the contrary, plasma estrone was only weakly positively associated with body mass index, systolic blood pressure, triglycerides and γGT while plasma estradiol was weakly negatively associated with age, fasting plasma glucose and total cholesterol.

The influence of aging on PTT is illustrated by Table 3. It shows a consistent PTT distribution decrement throughout the whole age range.

Table 1 Description of the two ethnic groups; mean (SEM in parentheses)

	Afro-Caribbean ($n = 186$)		Caucasian ($n = 1532$)		p
Age (years)	38.3	(0.7)	38.7	(0.3)	0.68
Body mass index (kg/m^2)	24.3	(0.2)	24.2	(0.1)	0.46
Subscapular skinfold (mm)	16.8	(0.6)	15.8	(0.2)	< 0.05
Systolic blood pressure (mmHg)	133	(1)	131	(1)	0.15
Diastolic blood pressure (mmHg)	79	(1)	78	(1)	0.12
Fasting plasma glucose (mg/dl)	92.5	(0.8)	93.4	(0.3)	0.29
2-h plasma glucose (mg/dl)	98.4	(2.2)	99.1	(0.7)	0.75
Fasting plasma insulin (μU/ml)	8.6	(0.3)	7.9	(0.1)	< 0.01
Total cholesterol (mg/dl)	225.0	(3.7)	229.6	(1.1)	0.23
Triglycerides (mg/dl)	83.9	(3.5)	98.5	(1.5)	< 0.001
γGT (U/l)	27.2	(1.9)	20.4	(0.5)	0.0001
Smokers (%)	24		37		< 0.001
Alcohol consumption (g/day)	23	(2)	21	(1)	0.29

Table 2 Correlation coefficients for sex hormones and covariates in men

	Estrone	Estradiol	Age	BMI*	Systolic BP*	G_0*	I_0*	G_2*	Cholesterol	Triglycerides	γGT
Total testosterone	0.11	0.26	-0.23	-0.29	-0.13	-0.19	-0.19	-0.20	-0.16	-0.23	-0.17
Estrone		0.47	0.01	0.06	0.07	-0.01	-0.01	0.01	0.00	0.05	0.12
Estradiol			-0.11	-0.03	0.02	-0.07	-0.03	-0.04	-0.09	-0.02	0.01

*BMI, body mass index; BP, blood pressure; G_0, fasting plasma glucose; I_0, fasting plasma insulin; G_2, 2-h plasma glucose $p < 0.05$ for $r \geq 0.05$, $p < 0.01$ for $r \geq 0.07$, $p < 0.001$ for $r \geq 0.09$

Table 3 Plasma total testosterone (ng/ml) distribution in men by 5-year age groups

Age (years)	n	Mean	SD	Median	1	5	10	90	95	99
								Percentiles		
<25	125	6.95	1.59	7.0	3.5	4.1	4.7	9.0	9.6	10.3
25–29	354	6.72	2.07	6.4	3.4	3.9	4.4	9.3	10.1	14.1
30–34	330	6.24	1.95	6.0	2.6	3.5	3.9	9.2	9.8	11.6
35–39	212	6.00	1.90	5.7	2.4	3.3	3.8	8.5	9.5	11.7
40–44	148	6.00	1.99	6.0	1.1	3.2	3.8	8.5	9.4	13.0
45–49	154	5.48	1.64	5.3	2.5	3.3	3.6	7.8	8.5	10.5
50–54	164	5.47	1.88	5.2	2.0	2.9	3.5	7.9	9.4	11.5
55–59	155	5.55	1.75	5.5	1.7	3.0	3.4	7.9	8.7	10.1
≥60	76	5.83	1.72	5.9	1.6	3.2	3.6	7.9	8.8	10.7
All	1718	6.12	1.95	5.9	2.4	3.4	3.8	8.7	9.6	11.4

The stepwise decrease of PTT with each decade of age was significant in univariate analysis ($p < 0.001$). In multiple linear regression analysis, using age as a continuous variable, adjustment for body mass index, subscapular skinfold, alcohol and tobacco consumption did not change the age-related decrement in PTT ($p < 0.0001$). The age-specific analyses were re-run after excluding the 111 men who reported a chronic disease and showed identical relationships, indicating that chronic diseases did not explain the decline in PTT level with age.

In order to have a better homogeneity of the sample, we focused on the Caucasians to look at the relationship of PTT with fasting plasma insulin in the cross-sectional study. In the Caucasian men, the correlation coefficient between PTT and fasting plasma insulin was −0.19 ($p < 0.0001$). In order to better describe the link between PTT and fasting plasma insulin, the crude and adjusted mean values of PTT were determined by categorical 2.5 µU/ml increments of fasting plasma insulin. As shown in Table 4, there was a significant stepwise decrease in PTT with increasing levels of insulin, even after adjusting for age ($p < 0.0001$). This decrease was reduced but remained graded and significant after simultaneous adjustment for age, body mass index, subscapular skinfold, tobacco and alcohol consumption, fasting and 2-h plasma glucose[2].

The design of the nested case–control study led to two groups of 30 men with contrasted PTT levels: 2.9 (0.1) vs. 5.9 (0.2) ng/ml ($p < 0.0001$), identical age and ethnic composition. In univariate analysis, the subjects with low PTT had significantly higher body mass index ($p < 0.01$), waist to hip ratio ($p < 0.001$), systolic blood pressure ($p < 0.05$), fasting and 2-h plasma glucose (both $p < 0.03$), fasting and 2-h plasma insulin (both $p < 0.0001$), triglycerides ($p < 0.01$), apo B lipoprotein ($p < 0.04$) and significantly lower HDL-cholesterol ($p < 0.02$) and apo A1 lipoprotein ($p < 0.04$) while total cholesterol was marginally higher in the low PTT group ($p = 0.07$). After adjustment for body mass index and waist to hip ratio, only fasting and 2-h plasma insulin remained statistically significantly higher in the low PTT group ($p < 0.04$ and $p < 0.004$ respectively). Measurements of plasma SHBG showed a significantly lower level in the low PTT men: 20.2 (1.3) vs. 37.9 (1.6) nmol/l ($p < 0.0001$), remaining significant after adjustment for body mass index and waist to hip ratio ($p < 0.0001$), while plasma bioavailable testosterone was not different between the two groups: 1897 (136) vs. 1975 (120) nmol/l.

Table 4 Mean plasma testosterone by level of fasting blood insulin before and after adjustment for age, obesity and tobacco and alcohol consumption

Plasma total testosterone (ng/ml)	*Fasting blood insulin* (μU/ml)				
	≤ 5 ($n = 229$)	5–7.5 ($n = 533$)	7.5–10 ($n = 339$)	> 10 ($n = 191$)	*p*
Unadjusted: mean	6.49	6.32	6.06	5.38	< 0.0001
(SD)	(1.97)	(1.89)	(1.85)	(1.60)	
Age adjusted	6.43	6.29	6.08	5.49	< 0.0001
Age, BMI* and skinfold adjusted	6.29	6.21	6.13	5.79	0.04
Age, BMI, skinfold, tobacco and alcohol adjusted	6.28	6.21	6.14	5.79	0.05

*BMI, body mass index

Tests were performed by analysis of variance; adjustments were performed by analysis of covariance using age, BMI, skinfold, tobacco and alcohol as continuous variables

The clinical trial comparing the effect of percutaneous administration of T, DHT and P gels over 3 months with six low PTT subjects by group showed a significant decrease of fasting plasma insulin under androgen treatment ($p = 0.02$ for the global comparison between the three groups; $p = 0.005$ between DHT and P; $p = 0.18$ between T and P) and a marginally significant decrease of systolic blood pressure under androgens ($p = 0.052$ for the global comparison; $p = 0.03$ between DHT and P; $p = 0.07$ between T and P). There was also a significant increase of hemoglobin and hematocrit induced by androgen treatments ($p = 0.03$ and 0.01 for the global comparison; $p = 0.08$ and 0.03 between DHT and P; $p = 0.006$ and 0.004 between T and P respectively) while the prostate-specific antigen (PSA) did not change. No difference was observed in lipid fractions, coagulation factors and fibrinolysis markers between the three groups of treatments.

DISCUSSION

Recruitment of this population of healthy adult men was occupation-based and consisted of volunteers. Therefore, the sample of the cross-sectional study cannot be considered as being representative of the total French adult male population, even if it is probably more representative of healthy people than most samples in other sex hormone studies, often based on patient populations consulting for diseases or sexual dysfunction.

Assaying sex hormones from a single sample for each individual certainly diminishes precision of the measurement but does not appear to constitute a limitation to the reliability of the data. In spite of the circadian variation of PTT[8,9], a single morning specimen has been found to be sufficiently reproducible for use in epidemiology[10]. Moreover, all the blood samples were drawn within a 1-h interval each morning, so no systematic bias should have been introduced by diurnal variation. In the same way, the well-known seasonal variation[9,11] would not explain the age-related changes in sex hormones because the period of examination was evenly distributed over the year and all the classes of age were equally represented over the seasons.

The ethnic differences for plasma estrogens confirm previous reports: higher levels of estradiol have been shown in male black compared to white Americans, in adolescents[12] as well as in adults[13]. Sex hormone differences could be involved in the racial differences for prostate cancer and cardiovascular diseases, both more frequent in black than in white Americans[14,15]. Estrogens have been suspected to promote prostate cancers in animal models[16] but the only prospective survey evaluating prediagnostic sex hormones as predictors of prostate cancer showed no significant association[17]. Additional epidemiological studies are needed to clarify the relationships between ethnic sex hormone differences and the risk of cardiovascular diseases and prostate cancer in ethnic groups.

Concerning the influence of aging on PTT, there was a small but steady stepwise decrement in total testosterone levels with age, beginning in young adult life. The slight increase of PTT over 55 years is probably explained by a 'healthy worker effect' in these older men who were still working and thus represented the healthier men in this age stratum. The similar trend for all percentiles and the remarkably similar standard deviations of testosterone between the decades of age indicate that the decrease of PTT with age is not explained by the contribution of a few outliers. The slope of the decline did not accelerate with the approach of middle age, nor was it explained by an age-related increment in known chronic diseases. These results do not support the thesis that the decline in PTT with age is due to disease, or that it begins late in life. These data are in accordance with previous studies[10,18–22] and a meta-analysis indicating a decrease of PTT with aging in men[23].

A negative PTT–insulin association has been observed in several other cross-sectional studies[24–27]. This negative relationship contradicts pharmacologic models which show the onset of impaired glucose tolerance and hyperinsulinemia when androgens are administered to male athletes[28] or to patients with aplastic anemia[29]. While cross-sectional observations in healthy men are closer to physiological conditions, they do not indicate the direction of the association between testosterone and insulin. The increase of plasma insulin with age is thought to be a physiologic marker of insulin resistance[30–34]. The early appearance of decreased insulin sensitivity, beginning as early as 30 years of age[30], could be the cause or the consequence of the parallel age-related decrease of PTT levels in men. The hypothesis that hyperinsulinemia is the initial event, preceding the decrease in plasma testosterone, is partly supported by an observed reduction of adrenal androgen production during hyperinsulinemic-euglycemic clamp in normal men but, unfortunately, testosterone evolution was not reported in the study concerned[35]. The observation of decreased free and total testosterone levels in older adult men with diabetes[36,37] can be considered as an indirect support to the hypothesis that hyperinsulinemia is the initial event. On the other hand, the hypothesis that a low level of plasma testosterone is the primary modification, followed by hyperinsulinemia, is suggested by the human model of Klinefelter's syndrome where dysgenesia-related hypogonadism is often accompanied by an impaired glucose tolerance, or even by diabetes, due to insulin resistance[38]. Moreover, the observation that lower levels of gonadal and adrenal androgens were associated with increasing levels of waist to hip ratio several years later in men of the Rancho Bernardo survey[39], suggests that low androgens promote abdominal obesity in men. As abdominal obesity is, in turn, a well-known predictor of hyperinsulinemia and diabetes[40], therefore it can be hypothesized that low PTT induces insulin resistance.

Although it remains an observational study, the data from our nested case–control survey reinforce the hypothesis of an association between PTT and insulin in men as PTT levels have been used as the only criterion to constitute the two groups. Nevertheless, the lower SHBG level in the PTT group, consistent with nearly identical bioavailable plasma testosterone in the two contrasted PTT groups, raises the issue of the respective roles of testosterone and SHBG in the association between sex hormones and insulin in men. If low SHBG has been shown to be an independent risk factor for non-insulin-dependent diabetes in women[41,42], such an association has not been demonstrated in men up to now[42]. [While the current paper was in press, an article has been published by Haffner, S. M. and co-workers showing that low SHBG levels predict the development of NIDDM in men[43]]. Our clinical trial indicates that, in men with low PTT, androgen treatment improves insulin sensitivity, as previously suggested in a few studies[44,45]. This effect of androgen therapy could be related to a reduction of abdominal adiposity[46,47], inducing in turn metabolic improvement[40,48]. Larger additional clinical trials are needed to confirm such an effect of androgen treatment and to evaluate, in the long-term, its cost to benefit ratio. So far, our observations suggest that in men with low PTT, androgen therapy may improve prediabetic conditions.

ACKNOWLEDGEMENTS

The authors are indebted to Laure Papoz, Dr Eveline Eschwege, Dr Marc Roger, Dr Evelyne Joubert, Professor Jean Rosa for scientific advice, to Edith Garat, Anne Forhan, Nadine Thibult, Liliane Simon and Patricia Grand for their excellent technical support, to the Fondation de Recherche en Hormonologie and to the Laboratoires Besins-Iscovesco for their invaluable contribution, and to the staff and the consultants of the Centre de Prévention Médicale des Télécommunications for their enthusiastic participation.

REFERENCES

1. Simon, D., Preziosi, P., Barrett-Connor, E., Roger, M., Saint-Paul, M., Nahoul, K. and Papoz, L. (1992). The influence of aging on plasma sex hormones in men: the Telecom Study. *Am. J. Epidemiol.*, **135**, 783–91

2. Simon, D., Preziosi, P., Barrett-Connor, E., Roger, M., Saint-Paul, M., Nahoul, K. and Papoz, L. (1992). Interrelation between plasma testosterone and plasma insulin in healthy adult men: the Telecom Study. *Diabetologia*, **35**, 173–7

3. Fontbonne, A., Papoz, L., Eschwège, E., Roger, M., Saint-Paul, M. and Simon, D. (1992). Features of insulin-resistance syndrome in men from French Caribbean islands. The Telecom Study. *Diabetes*, **41**, 1385–9

4. Scholler, R., Nahoul, K., Castanier, M., Rotman, J. and Salat-Baroux, J. (1984). Testicular secretion of conjugated and unconjugated steroids in normal adults and in patients with varicocele. *J. Steroid Biochem.*, **20**, 203–15

5. Castanier, M. and Scholler, R. (1970). Dosage radio-immunologique de l'estrone et de l'estradiol-17β plasmatiques (in French). *Compte Rendu Acad Sci (III)*, **271**, 1787–9

6. Nahoul, K. and Roger, M. (1990). Age related decline of plasma bioavailable testosterone in adult men. *J. Steroid Biochem.*, **35**, 293–9

7. Simon, D., Fagot, A., Nahoul, K. and Papoz, L. (1993). Racial differences in the level of sex hormones in healthy adult men: the Telecom Study. Oral communication at the 13th International Epidemiology Association meeting, Sydney (Australia), 1993

8. De Lacerda, L., Kowarski, A., Johanson, A. J., Athanasiou, R. B. and Migeon, C. J. (1973). Integrated concentration and circadian variation of plasma testosterone in normal men. *J. Clin. Endocrinol. Metab.*, **37**, 366–71

9. Reinberg, A., Lagoguey, M., Chauffournier, J. M. and Cesselin, F. (1975). Circannual and circadian rhythms in plasma testosterone in five healthy young Parisian males. *Acta Endocrinol.*, **80**, 732–43

10. Dai, W. S., Kuller, L. H., La Porte, R. E., Gutai, J. P., Falvo-Gerard, L. and Caggiula, A. (1981). The epidemiology of plasma testosterone levels in middle-aged men. *Am. J. Epidemiol.*, **114**, 804–16

11. Smals, A. G. H., Kloppenborg, P. W. C. and Benraad, T. J. (1976). Circannual cycle in plasma testosterone levels in man. *J. Clin. Endocrinol. Metab.*, **42**, 979–82

12. Srinivasan, S. R., Freedman, D. S., Sundaram, G. S., Webber, L. S. and Berenson, G. S. (1986). Racial (black–white) comparisons of the relationship of levels of endogenous sex hormones to serum lipoproteins during male adolescence: the Bogalusa Heart Study. *Circulation*, **74**, 1226–34

13. Hughes, G. S., Mathur, R. S. and Margolius, H. S. (1989). Sex steroid hormones are altered in essential hypertension. *J. Hypertens.*, **7**, 181–7

14. Ross, R., Bernstein, L., Judd, H., Hanisch, R., Pike, M. and Henderson, B. (1986). Serum testosterone levels in healthy young black and white men. *J. Natl. Cancer Inst.*, **76**, 45–8

15. Cooper, R. S. and Ford, E. (1992). Comparability of risk factors for coronary heart disease among Blacks and Whites in the NHANES-I epidemiologic follow-up study. *Ann. Epidemiol.*, **2**, 637–45

16. Santti, R., Pylkkänen, L., Newbold, R. and McLachlan, J. A. (1990). Developmental oestrogenization and prostatic neoplasia. *Int. J. Androl.*, **13**, 77–80

17. Nomura, A., Heilbrun, L. K., Stemmermann, G. N. and Judd, H. L. (1988). Prediagnostic serum hormones and the risk of prostate cancer. *Cancer Res.*, **48**, 3515–17

18. Vermeulen, A. (1967). Testosterone excretion, plasma levels and testosterone production in health and disease. In Tamm, J. (ed.) *Testosterone* pp. 170–5. (New York: Hatner)

19. Pirke, K. N. and Doerr, P. (1973). Age-related changes and interrelationships between plasma testosterone, estradiol and testosterone-binding globulin in normal adult males. *Acta Endocrinol.*, **74**, 792–800

20. Davidson, J. M., Chen, J. J., Crapo, L., Gray, G. D., Greenleaf, W. J. and Catania, J. A. (1983). Hormonal changes and sexual function in aging men. *J. Clin. Endocrinol. Metab.*, **57**, 71–7

21. Rudman, D., Mattson, D. E., Nagraj, H. S., Feller, A. G., Jackson, D. L. and Rudman, I. W. (1988). Plasma testosterone in nursing home men. *J. Clin. Epidemiol.*, **41**, 231–6

22. Deslypere, J. P. and Vermeulen, A. (1984). Leydig cell function in normal men: effect of age, life-style, residence, diet and activity. *J. Clin. Endocrinol. Metab.*, **59**, 955–62

23. Gray, A., Berlin, J. A., McKinlay, J. B. and Longcope, C. (1991). An examination of research design effects on the association of testosterone and male aging: results of a meta-analysis. *J. Clin. Epidemiol.*, **44**, 671–84

24. Lichtenstein, M. J., Yarnell, J. W. G., Elwood, P. C., Beswick, A. D., Sweetnam, P. M., Marks, V., Teale, D. and Riad-Fahmy, D. (1987). Sex hormones, insulin, lipids, and prevalent ischemic heart disease. *Am. J. Epidemiol.*, **126**, 647–57

25. Phillips, G. B. (1993). Relationship between serum sex hormones and the glucose–insulin–lipid defect in men with obesity. *Metabolism*, **42**, 116–20

26. Haffner, S. M., Valdez, R. A., Mykkänen, L., Stern, M. P. and Katz, M. S. (1994). Decreased testosterone and dehydroepiandrosterone sulfate concentrations are associated with increased insulin and glucose concentrations in nondiabetic men. *Metabolism*, **43**, 599–603

27. Haffner, S. M., Karhapää, P., Mykkänen, L. and Laakso, M. (1994). Insulin resistance, body fat distribution, and sex hormones in men. *Diabetes*, **43**, 212–19

28. Cohen, J. C. and Hickman, R. (1987). Insulin resistance and diminished glucose tolerance in powerlifters ingesting anabolic steroids. *J. Clin. Endocrinol. Metab.*, **64**, 960–3

29. Woodward, T. L., Burghen, G. A., Kitabchi, A. E. and Wilimas, J. A. (1981). Glucose intolerance and insulin resistance in anaplastic anemia treated with oxymetholone. *J. Clin. Endocrinol. Metab.*, **53**, 905–8

30. De Fronzo, R. A. (1981). Glucose intolerance and aging. *Diabetes Care*, **4**, 493–501

31. Fink, R. I., Kolterman, O. G., Griffin, J. and Olefsky, J. M. (1983). Mechanisms of insulin resistance in aging. *J. Clin. Invest.*, **71**, 1523–35

32. Rowe, J. W., Minaker, K. L., Pallotta, J. A. and Flier, J. S. (1983). Characterization of the insulin resistance of aging. *J. Clin. Invest.*, **71**, 1581–7

33. Bolinder, J., Östman, J. and Arner, P. (1983). Influence of aging on insulin receptor binding and metabolic effects of insulin on human adipose tissue. *Diabetes*, **32**, 959–64

34. Jackson, R. A. (1990). Mechanisms of age-related glucose intolerance. *Diabetes Care*, **13** (Suppl. 2), 9–19

35. Nestler, J. E., Usiskin, K. S., Barlascini, C. O., Welty, D. F., Clore, J. N. and Blackward, W. G. (1989). Suppression of serum dehydroepiandrosterone sulfate levels by insulin: an evaluation of possible mechanisms. *J. Clin. Endocrinol. Metab.*, **69**, 1040–6

36. Ando, S., Rubens, R. and Rottiers, R. (1984). Androgen plasma levels in male diabetics. *J. Endocrinol. Invest.*, **7**, 21–4

37. Barrett-Connor, E., Khaw, K. T. and Yen, S. S. C. (1990). Endogenous sex hormone levels in older adult men with diabetes mellitus. *Am. J. Epidemiol.*, **132**, 895–901

38. Geffner, M. E., Kaplan, S. A., Bersch, N., Lippe, B. M., Bergman, R. N. and Golde, D. W. (1985). Insulin resistance in Klinefelter's syndrome. *Clin. Res.*, **33**, 128A

39. Khaw, K. T. and Barrett-Connor, E. (1992). Lower endogenous androgens predict central adiposity in men. *Ann. Epidemiol.*, **2**, 675–82

40. Larsson, B., Svärdsudd, K., Welin, L., Wilhelmsen, L., Björntorp, P. and Tibblin, G. (1984). Abdominal adipose tissue distribution, obesity, and risk of cardiovascular disease and death: 13 years follow-up of the participants in the study of men born in 1913. *Br. Med. J.*, **288**, 1401–4

41. Lindstedt, G., Lundberg, P. A., Lapidus, L., Lundgren, H., Bengtsson, C. and Björntorp, P. (1991). Low sex-hormone-binding globulin concentration as independent risk factor for development of NIDDM. *Diabetes*, **40**, 123–8

42. Haffner, S. M., Valdez, R. A., Morales, P. A., Hazuda, H. P. and Stern, M. P. (1993). Decreased sex-hormone-binding globulin predicts non-insulin-dependent diabetes mellitus in women but not in men. *J. Clin. Endocrinol. Metab.*, **77**, 56–60

43. Haffner, S. M., Shaten, J., Stern, M. P., Smit, G. D. and Kuller, L. (1996). Low levels of sex hormone-binding globulin and testosterone predict the development of non-insulin-dependent diabetes mellitus in men. *Am. J. Epidemiol.*, **143**, 889–97

44. Marin, P., Holmäng, S., Jönsson, L., Sjöström, L., Kvist, H., Holm, G., Lindstedt, G. and Björntorp, P. (1992). The effects of testosterone treatment on body composition and metabolism in middle-aged obese men. *Int. J. Obesity*, **16**, 991–7

45. Marin, P., Holmäng, S., Gustafsson, C., Jönsson, L., Kvist, H., Elander, A., Eldh, J., Sjöström, L., Holm, G. and Björntorp, P. (1993). Androgen treatment of abdominally obese men. *Obesity Res.*, **1**, 245–51

46. Rebuffé-Scrive, M., Marin, P. and Björntorp, P. (1991). Effect of testosterone on abdominal adipose tissue in men. *Int. J. Obesity*, **15**, 791–5

47. Marin, P., Oden, B. and Björntorp, P. (1995). Assimilation and mobilization of triglycerides in subcutaneous abdominal and femoral adipose tissue *in vivo* in men: effects of androgens. *J. Clin. Endocrinol. Metab.*, **80**, 239–43

48. Björntorp, P. (1991). Metabolic implications of body fat distribution. *Diabetes Care*, **14**, 1132–43

DISCUSSION

Orwoll: With reference to the paradoxes set forward by Dr Haffner, there is another paradox that I would like you to comment on. In small interventional studies in which androgens are manipulated, it looks like testosterone levels are inversely related to HDL levels. For instance, in Bremner's work in young men[1], GnRH agonists lowered testosterone levels dramatically, whereas they clearly increased HDL levels. This seems to be the opposite of what you would predict from cross-sectional studies. Can you explain that paradox?

Haffner: In the studies you mention, you are dealing with a central effect, of giving GnRH as opposed to giving peripheral testosterone. The mode of how you are affecting testosterone levels is likely to have a major influence and I would speculate that this explains the paradox. Nevertheless, there have been only a few studies which have really worked at HDL cholesterol levels in men specifically, and they have been relatively small. So I am not sure that we actually have enough data to settle this issue.

Simon: I would like to come back to SHBG as a predictor for NIDDM. In conclusion, you say that low testosterone and low SHBG predict NIDDM. In women this was demonstrated by Lindstedt and co-workers in Göteborg[2]. But are your own data[3] not the opposite for men?

Haffner: Well no. Both in men and women low SHBG predicts the incidence of diabetes. Lindstedt only measured SHBG, but in our study in which we measured more than SHBG, we also saw that low testosterone levels predict diabetes[4]. The SHBG data are thus actually consistent in men and women.

Kaufman: There is some good indication that estrogen in post-menopausal women may be protective for cardiovascular disease. We do not know the exact mechanism, but direct vascular effects seem to be involved in part. Is it possible that the lack of association between cardiovascular risk factors and actual cardiovascular disease may be due to the fact that, apart from their effects on risk factors, sex steroids also have direct vascular effects?

Haffner: This is an extremely important issue: First of all, there is a very large set of observational data, suggesting that estrogen replacement decreases cardiovascular disease in women. There are no clinical trial data yet, although there is a megatrial going on in the United States (the Women's Health Initiative). Estrogen effects on cardiovascular disease are likely to be mediated both by effects on risk factors for atherosclerosis and by effects on vascular reactivity. It is not clear which of the two is more important at this point. It is possible that testosterone replacement also affects vascular reactivity. However, if it is true that there is relatively little effect of testosterone in terms of prediction of cardiovascular disease, whereas it seems that there are generally beneficial effects on traditional cardiovascular risk factors, then one might postulate that testosterone replacement may have adverse effects on cardiovascular reactivity. Nevertheless, research must reveal whether this is indeed the case.

Bergink: You predicted that testosterone replacement therapy will not be cardioprotective in elderly men. This prediction is based on surrogate markers under pathological conditions and then extrapolated to normal treatment conditions. Taking such an extrapolation into consideration, do you still maintain the prediction that androgen administration will not be cardioprotective?

Haffner: There are two issues. First, I think androgen administration is very unlikely to be helpful in the general male population. This is one of the points I want to make: if there is going to be an effect, it is going to be in men who in fact have very low androgen levels. In my opinion, however, there are no data to suggest that administration is going to reduce car-diovascular disease. One can criticize the available studies because these prospective studies are looking at the overall population, and not specifically at hypogonadal men, but I think that the burden of proof is showing. However, I also think that there is at least a reasonable possibility that administration may reduce the risk of diabetes in high risk men, and that might be a much more fruitful way to look at the issue.

REFERENCES

1. Bagatell, C. J., Knopp, R. H., Vale, W. W., Rivier, J. E., Bremner, W. J. (1992). Physiologic testosterone levels in normal men suppress high-density lipoprotein cholesterol levels. *Ann. Intern. Med.*, **116**, 967–73
2. Lindstedt, G., Lundberg, P. A., Lapidus, L., Lundgren, H., Bengtsson, C., Björntorp, P. (1991). Low sex hormone binding globulin concentration as independent risk factors for development of NIDDM: 12-yr. follow-up of population study of women in Gothenburg, Sweden. *Diabetes*, **40**, 123–8
3. Haffner, S. M., Valdez, R. A., Morales, P. A., Hazuda, H. P., Stern, M. P. (1993). Decreased sex hormone binding globulin predicts non-insulin dependent diabetes mellitus in women but not in men. *J. Clin. Endocrinol. Metab.*, **77**, 56–60
4. Haffner, S. M., Shaten, J., Stern, M. P., Smith, G. D., Kuller, L. K. (1996). Low levels of sex hormone binding globulin and testosterone predict the incidence of NIDDM in men. The Multiple Risk Factor Intervention Trial. *Am. J. Epidemiol.*, **143**, 889–97

6

Muscle mass and fat distribution in relation to androgens

C. Longcope

An androgen is generally considered to be a hormone related to the male reproductive tract. Indeed, an androgen 'may be defined as a substance which is capable of stimulating male secondary sex characteristics'[1]. An androgen may have other actions, and it has long been known that androgens influence muscle mass and more recently that androgens may influence fat distribution.

Muscle mass is decreased in hypogonadal males as compared to normals, and the administration of androgens to such individuals will result in an increase in muscle mass. The increase was originally thought to be universal for all skeletal muscles[2]. Subsequently, studies in castrated rats indicated that the muscles of the thigh were not fully restored with androgen administration. In castrated males of several species the administration of androgens results in an increase in the mass of the muscles of the upper back, shoulders and head[2-4], with lesser stimulation to other muscles. The exact reason that those muscles are specifically stimulated by androgens is unclear.

Androgens increase muscle mass by increasing the size of the muscle fibers[5]. Following castration there is a slow dissolution of the myofibrils and loss of sarcoplasm. With testosterone replacement the myofibrils increase in number leading to an enlargement of the muscle cell and mass. This action appears to be receptor mediated and androgen receptors have been characterized in muscle[6,7] although their importance has been questioned[8]. The receptors in muscle appear to be similar to those in the prostate[6] with a K_D of $2 \times 10^{-7} - 5 \times 10^{-7}$. Although the receptor binds both testosterone and dihydrotestosterone, testosterone is the major androgen in muscle. The enzyme responsible for the conversion of testosterone to dihydrotestosterone, 5α-reductase, is at very low levels in muscle[9]. Therefore, little dihydrotestosterone is formed in the muscle, and the circulating levels of dihydrotestosterone are too low to have appreciable effect.

The action of androgens on muscle is categorized as one of the anabolic effects of androgens in comparison to the androgenic action on reproductive tissues. The effect of androgens on nitrogen balance falls into the same category, and in hypogonadal men the administration of testosterone will lead to a positive nitrogen balance[2,3,10]. Because of this effect of testosterone, androgens have been administered to chronically ill, malnourished people in an attempt to restore weight and muscle mass[11]. While a positive nitrogen balance is achieved initially such an effect is short-lived, weeks to months, and not sustained[3]. When the androgen is stopped there may be an accelerated loss of nitrogen so little benefit results. It should be noted that the effect of androgen on nitrogen balance requires a threshold level of insulin for the stimulatory effect[2].

The role of androgen as a means of increasing athletic prowess in men is a subject of debate. Many individuals in purely strength–related endeavours, such as weight-lifting, and weight-throwing, have been using large amounts of androgens to increase muscle mass and performance. It is not completely clear whether androgen administration will increase strength in such individuals and the literature reflects the uncertainty[12]. Although there are reports relating androgen administration to improved strength, these reports are often poorly designed and definite conclusions are hard to draw. Those that are well designed with good controls usually show little or no benefit from the androgen. However, the amounts of androgens used in the latter studies are generally well below those used by athletes. Whether the greater dose does have an effect is not clear and may never be settled[12]. Confounding the issue as to the effect of androgens are the roles of training and nutrition for both can result in increased performance independent of androgen use.

In women the administration of androgens can result in increasing muscle mass and strength. The difference in the result in men and women is that the latter start from a baseline androgen level similar to that in hypogonadal males. The increase in testosterone and androgens will thus result in an increase in performance.

While the effect of androgens on muscle mass has been appreciated for a relatively long time, that androgens may effect fat distribution is a relatively new concept. The possible role of androgens on adipose tissue distribution was emphasized by studies of Faggiano and co-workers[13] noting that in most men the neck, shoulders and abdomen are the sites of adiposity while in many women the adipose tissue is deposited in the buttocks and thighs.

This relationship between androgens and adipose tissue distribution was explored further by Kissebah and Peiris[14] and Björntorp[15]. These investigators noted that upper body obesity in women was associated with a number of metabolic abnormalities, e.g. insulin resistance and impaired glucose tolerance. Evans and colleagues[16,17] reported that in premenopausal women the waist to hip ratio was inversely correlated with the levels of sex hormone-binding globulin and directly correlated with percentage free testosterone. They found no relationship between total testosterone, androstenedione, or dehydroepiandrosterone and waist to hip ratio.

Unfortunately, they did not give data on the concentration of free testosterone, and it is possible that the concentration of free testosterone would not have correlated with the waist to hip ratio. However, Kirschner and colleagues[18] did report that women with upper body obesity (waist to hip ratio > 0.85) had higher testosterone production rates than did women with lower body obesity (waist to hip ratio < 0.75). It is not possible, from their data, to determine whether the higher testosterone production was a cause of or a result of the higher waist to hip ratio. Barbieri and Hornstein[19] reported that insulin will increase ovarian androgen production. Thus, it is possible that increased abdominal fat results in higher insulin levels[14,15] that in turn lead to more androgen production.

Adipocytes differ in their histological and biochemical characteristics depending on whether they are obtained from the abdominal or femoral areas[20]. Abdominal fat is made up primarily of large adipocytes that aromatize less well than the smaller adipocytes that make up most of the femoral adipose tissue[20]. Lipoprotein lipase activity in abdominal fat is decreased by androgen administration whereas lipoprotein lipase activity in femoral fat is increased by androgen administration[21]. With testosterone administration triglyceride uptake is greater in the abdominal area than in the femoral area[21].

Adipose tissue metabolizes androgens[22] and there is a direct correlation between body weight and the metabolic clearance rate of androgens[23]. Kirschner and co-workers[18] noted that the women with upper body obesity had a significantly greater metabolic clearance rate of testosterone than the women with lower body obesity even though the mean weights of the two groups were not different.

As noted, the distribution of adipose tissue tends to differ in women as compared to men, and a high waist to hip ratio is associated with increased circulating androgens in women[15,17].

Obesity in men is usually associated with low levels of testosterone and sex hormone-binding globulin[24]. In hypogonadal men with low levels of androgens or obese men, the administration of androgens results in a decrease in abdominal or visceral fat[25]. However, in eugonadal men there is usually a reduction in fat around the hips and an increase in upper body fat, and, as noted previously, in muscle mass.

As men age there is a decrease in muscle mass and some increase in adipose tissue, especially in the abdominal and upper body regions[26,27]. Also associated with aging is a decrease in the circulating levels of testosterone[28-30] that, in many men, reach the level seen in young hypogonadal men. While androgen administration will increase muscle mass and alter fat distribution in young hypogonadal men, it has not been rigorously shown that androgen administration to older men with low testosterone levels will have the same effects. Tenover[31] did note that there was an increase in muscle mass in older men with low testosterone levels who were given testosterone enanthate, 100 mg weekly for 3 months. Body weight was also increased but neither per cent body fat nor waist to hip ratio was altered by treatment. Thus, while it appears that testosterone administration does not have quite as much effect on muscle mass and fat distribution in older men as in young hypogonadal men, further studies are needed.

Although 5α-reductase activity is present in fat tissue, it is at a low level and testosterone is the major androgen. Androgen receptors have been found in adipose tissue[32]. The contributions of the other C-19 steroids, e.g. androstenedione and dehydroepiandrosterone, are small and probably are due to their conversions to the more active androgen testosterone.

In summary, androgens increase muscle mass but this effect is most marked in individuals with low androgen levels, e.g. women and hypogonadal males. The effects in older men with low levels of testosterone are still being evaluated. Androgens also affect adipose tissue distribution, but not necessarily mass, generally leading to a greater disposition of fat in the upper body, although in situations of low testosterone levels the reverse has been observed.

REFERENCES

1. Dorfman, R. I. and Shipley, R. A. (1956). General concepts. In *Androgens*, pp. 3–4. (New York: John Wiley & Sons)

2. Russell, J. A. and Wilhelmi, A. E. (1960). Endocrines and muscle. In Bourne, G. H. (ed.) *The Structure and Function of Muscle*, pp. 146–50 (New York: Academic Press)

3. Wilson, J. D. and Griffin, J. E. (1980). The use and misuse of androgens. *Metabolism*, **29**, 1278–95

4. Gooren, L. J. G. and Polderman, K. H. (1990). Safety aspects of androgen therapy. In Nieschlag, E. and Behre, H. M. (eds.) *Testosterone: Action Deficiency Substitution*, pp. 182–203. (Berlin: Springer-Verlag)

5. Venable, J. H. (1995). Morphology of the cells of normal testosterone-deprived and testosterone-stimulated levator ani muscles. *Am. J. Anat.*, **119**, 271–302

6. Michel, G. and Baulieu, E. E. (1980). Androgen receptor in rat skeletal muscle: characterization and physiological variations. *Endocrinology*, **107**, 2088–98

7. Anselmo, V. C., Buchanan, J. M. and Sheridan, P. J. (1980). The heart is a target organ for androgen. *Science*, **207**, 775–7

8. Menon, M., Tananis, C. E., Hicks, L. L., Hawkins, E. F. and McLoughlin, M. G. (1978). Characterization of the binding of a potent synthetic androgen, methyltrienolone, to human tissues. *J. Clin. Invest.*, **61**, 150–62

9. Wilson, J. D. and Gloyna, R. E. (1970). The intranuclear metabolism of testosterone in the accessory organs of reproduction. *Recent Prog. Horm. Res.*, **26**, 309

10. Kochakian, C. D. (1990). History of anabolic–androgenic steroids. *NIDA Research Monogr.*, **102**, 29–59

11. Tweedle, D., Walton, C. and Johnston, I. D. A. (1973). The effect of an anabolic steroid on postoperative nitrogen balance. *Br. J. Clin. Pract.*, **27**, 130–2

12. Wilson, J. D. (1988). Androgen abuse by atheletes. *Endocr. Rev.*, **9**, 181–99

13. Faggiano, M., Carella, C., Criscuolo, T., Jaquet, P. and Minozzi, M. (1974). Massive obesity and thyroid function. In Vague, J. and Boyer, J. (eds.) *The Regulation of the Adipose Tissue Mass*, p. 295. (New York: American Elsevier)

14. Kissebah, A. H. and Peiris, A. N. (1989). Biology of regional body fat distribution: relationship to non-insulin-dependent diabetes mellitus. *Diabetes Metab. Rev.*, **5**, 95–109

15. Björntorp, P. (1988). The association between obesity, adipose tissue distribution, and disease. *Acta Med. Scand.*, **723**, 121–34

16. Evans, D. J., Hoffman, R. G., Kalkhoff, R. K. and Kissebah, A. H. (1983). Relationship of androgenic activity to body fat topography, fat cell morphology, and metabolic aberrations in premenopausal women. *J. Clin. Endocrinol. Metab.*, **57**, 304–10

17. Evans, D. J., Barth, J. H. and Burke, C. W. (1988). Body fat topography in women with androgen excess. *Int. J. Obesity*, **12**, 157–62

18. Kirschner, M. A., Samojlik, E., Drejka, M., Szmal, E., Schneider, G. and Ertel, N. (1990). Androgen–estrogen metabolism in women with upper body *versus* lower body obesity. *J. Clin. Endocrinol. Metab.*, **70**, 473–9

19. Barbieri, R. L. and Hornstein, M. D. (1988). Hyperinsulinemia and ovarian hyperandrogenism. *Endocrinol. Metab. Clin. North Am.*, **17**, 685–703

20. Killinger, D. W., Perel, E., Daniilescu, D., Kharlip, L. and Lindsay, W. R. N. (1990). Influence of adipose tissue distribution on the biological activity of androgens. *Ann. N. Y. Acad. Sci.*, **595**, 199–211

21. Marin, P., Oden, B. and Bjorntorp, P. (1995). Assimilation and mobilization of triglycerides in subcutaneous abdominal and femoral adipose tissue *in vivo* in men: effects of androgens. *J. Clin. Endocrinol. Metab.*, **80**, 239–43

22. Longcope, C., Pratt, J. H., Schneider, S. H. and Fineberg, S. E. (1976). The *in vivo* metabolism of androgens by muscle and adipose tissue of normal men. *Steroids*, **28**, 521–33

23. Longcope, C. and Baker, S. (1993). Androgen and estrogen dynamics: relationships with age, weight, and menopausal status. *J. Clin. Endocrinol. Metab.*, **76**, 601–4

24. Tchernof, A., Despres, J. P., Belanger, A., Dupont, A., Prudhomme, D., Moorjani, S., Lupien, P. J. and Labrie, F. (1995). Reduced testosterone and adrenal C19 steroid levels in obese men. *Metabolism*, **44**, 513–9

25. Marin, P., Holmang, S., Jonsson, L., Sjostrom, L., Kvist, H., Holm, G. and Lindstedt, G. (1992). The effects of testosterone treatment on body composition and metabolism in middle-aged obese men. *Int. J. Obesity.*, **16**, 991–7

26. Tenover, J. S. (1994). Androgen administration to aging men. *Endocrinol. Metab. Clin. North Am.*, **23**, 877–92

27. Swerdloff, R. S. and Wang, C. (1993). Androgens and aging in men. *Exp. Gerontol.*, **28**, 435–46

28. Vermeulen, A., Rubens, R. and Verdonck, L. (1972). Testosterone secretion and metabolism in male senescence. *J. Clin. Endocrinol. Metab.*, **34**, 730–5

29. Deslypere, J. P. and Vermeulen, A. (1984). Leydig cell function in normal men: effect of age, lifestyle, residence, diet, and activity. *J. Clin. Endocrinol. Metab.*, **59**, 955–62

30. Gray, A., Feldman, H. A., McKinlay, J. B. and Longcope, C. (1991). Age, disease, and changing sex hormone levels in middle-aged men: results of the Massachusetts male aging study. *J. Clin. Endocrinol. Metab.*, **73**, 1016–25

31. Tenover, J. S. (1992). Effects of testosterone supplementation in the aging male. *J. Clin. Endocrinol. Metab.*, **75**, 1092–8

32. De Pergola, G., Xu, X., Yang, S., Giorgino, R. and Björntorp, P. (1990). Up-regulation of androgen receptor binding in male rat fat pad adipose precursor cells exposed to testosterone: study in a whole cell assay system. *J. Steroid Biochem. Mol. Biol.*, **37**, 553–8

DISCUSSION

Bain: The popular press and the public at large believe that the administration of androgens, particularly in large doses to normal men, will increase muscle mass and strength, and you have pointed out that, in fact, there is precious little evidence for this. When you talk to athletes they say they do not care what the scientists say. They are sure that it helps them and makes their body more beautiful and their muscles stronger. You quoted some studies showing that there is no change in muscle size, but you did show evidence that there was increased protein synthesis in muscles. I wonder if the athletes are possibly not entirely wrong, although we have precious little evidence, that their strength might be increased. The protein synthesis mechanism might then, to some degree, account for that.

Longcope: This is an issue that will be very hard to ever determine by research, because I suspect no institutional review board would allow a researcher to administer androgens so that levels of 50 ng/ml in men or 15 ng/ml in women are reached. An individual that I personally studied some years ago stated that he just knew that the steroids were helping him. He felt, when he went into the weight room, that he could bite the iron bar in two if his testosterone levels were 50 ng/ml. If they were much lower, he said he could not do anything. I think that training has an effect and certainly in some of the studies the effects of androgens were more marked in those who trained rather than in those who did not train. Nevertheless, it is true that it is almost impossible to convince somebody who has been on steroids that he can do as well without them.

[**Dr Bain** adds: Some months after the Workshop, a study addressing this issue was published in the New England Journal of Medicine[1]. Men treated with 600 mg testosterone enanthate intramuscularly, weekly, for 10 weeks, even if they did not exercise, had greater increases in muscle size and strength when compared to a placebo-treated group. Exercise enhanced the increase in muscle size and strength even further. There was no change in mood or behavior.]

Gooren: We did a follow-up study on muscle mass in female to male transsexuals on high doses of testosterone. It was the conventional dose used in hypogonadal men, that is, 250 mg of testosterone esters per 2 weeks. An increase in muscle mass was observed of 1.5 kilos of lean body mass, but only in the first year and thereafter it stopped.

Schröder: Has anybody ever looked at the presence or absence of androgen receptors in the myocardium?

Longcope: Yes, there was a study on the topic[2] and they did find that there was an effect of androgens on the cardiac muscle, and I believe that androgen receptors were present.

REFERENCES

1. Bhasin, S., Storer, T. W., Berman, N., Callegari, C., Clevenger, B., Phillips, J., Bunnell, T. J., Tricker, R., Shirazi, A. and Casaburi, R. (1996). The effects of supraphysiologic doses of testosterone on muscle size and strength in normal men. *New Engl. J. Med.*, **335**, 1–7
2. McGill, H. C. Jr., Anselmo, V. C., Buchanan, J. M. and Sheridan, P. J. (1980). The heart is a target organ for androgen. *Science* **207**, 775–7

7

Androgens and sexual function in men

R. C. Schiavi

The essential role of androgens for sustaining sexual behavior in subhuman mammals has been known for well over half a century. It was not until the late 1970s, however, that the systematic investigation of testicular androgens in human sexuality was initiated. Although much remains to be known, a substantial body of information on androgen effects on male sexual behavior has recently become available. Findings from studies on androgen administration to typical hypogonadal men (e.g. as a result of congenital dysgenesis) will be summarized prior to considering the role of androgens in the sexual functioning of sexually normal as well as dysfunctional eugonadal individuals. This overview will provide a background for discussing research from our laboratories on the relation between pituitary and gonadal hormones and sexual function in aging men and on the sexual effects of androgen supplementation.

STUDIES ON HYPOGONADAL MEN

Several controlled studies have clearly shown that androgen replacement therapy restores sexual function in hypogonadal patients. Testosterone administration to men with subnormal levels of circulating androgens stimulates their sexual interest, increases the frequency of sexual activity and re-establishes ejaculatory capacity[1-3]. Dose–response relationships have been demonstrated for frequency of sexual thoughts, sexual excitement accompanying such thoughts and total number of erections per week[4,5]. The degree, duration and nature of the changes in circulating androgens following exogenous testosterone have varied widely within and across studies depending on androgen preparations, routes of administration and dosage schedules[6]. There still is conflicting information as to how androgen influences male sexual function and the extent to which aromatization to estrogen or 5α-reduction to dihydrotestosterone mediates its action in human sexuality.

Recent studies have begun to clarify the degree to which testosterone may influence erectile function independently from its primary effect on sexual interest. The mechanism of behavioral action of androgens has been explored in hypogonadal men by recording sleep-related nocturnal erections and erectile responses to control presentation of erotic stimuli. Sleep-related erections which are usually impaired in hypogonadal men are restored by testosterone independent of changes in sleep time and rapid eye movement (REM) activity[4]. In contrast, erections induced in response to visual erotic stimuli in non-treated hypogonadal men do not appear to differ from those of normal subjects and are not affected by androgen replacement[3-5,7]. These findings have led to the conclusion that androgens are more important to sustain sexual interest and sleep-related erections than they are to maintain erectile responses to external stimuli. The erectile difficulties observed in hypogonadal men are interpreted as secondary to a decrease in sexual desire. However, while Bancroft and Wu postulate that the effect of androgen on sexual interest is *via* central cognitive processes and is possibly fantasy mediated[7], Davidson and colleagues have proposed that the primary effect is peripheral, acting through possible changes in genital sensitivity and the pleasurable enhancement of erectile activity[6]. Support for these alternative hypotheses rests on scarce evidence, mostly from short-term androgen studies on erectile responses to experimentally induced erotic stimulation.

A review of studies on hypogonadal men emphasizes the value of controlled prospective investigations of androgen therapy, the importance of carefully differentiating and measuring sexual interest, arousal, activity and satisfaction, and the significant contribution made by the objective recording of erectile responses in the laboratory. It suggests that the measuring of nocturnal penile tumescence (NPT) provides an avenue to explore central, androgen-dependent processes possibly involved in behavior. The above studies have been limited, however, by the small number of subjects and their clinical heterogeneity, and the narrow scope of the inquiry on affective and psychological responses. Some[1,4], but not all studies[2,5], have found significant positive androgen effects on the well-being and mood of hypogonadal men. The extent to which changes in mood may have influenced sexual interest and activity is not known.

STUDIES ON EUGONADAL MEN

There are wide individual variations in circulating testosterone among healthy adult men but there is no convincing evidence that testosterone levels, within the normal range, account for differences in sexual behavior. It is assumed that men have plasma androgens at concentrations substantially higher than the threshold levels required for behavioral activation. Male sexuality, however, encompasses behavioral and psychophysiologic components that may vary in degree of androgen dependency. Degree and latency of erectile responses to erotic films in eugonadal, sexually non-dysfunctional men correlates with endogenous testosterone levels suggesting that testosterone, within the normal range, may influence some aspects of erectile function[8,9]. It has also been found[10] that the administration of testosterone to normal subjects increased the rigidity of NPT episodes. It should be noted that changes in erectile capacity, measured in the laboratory, may not be reflected in erectile function in interpersonal situations.

Controlled studies on the circulating hormonal levels required to sustain sexual behavior in healthy, sexually non-dysfunctional, men are extremely few. Bagatell and co-workers[11] examined the effect of pharmacological changes in endogenous testosterone on normal young volunteers. Induction of an acute, reversible, gonadal steroid deficiency with a gonadotropin-releasing hormone (GnRH) antagonist resulted in marked decrements in frequency of sexual desire, sexual fantasies and coital activity 4–6 weeks after treatment onset. Sexual function was restored following termination of treatment in parallel with the normalization of circulating testosterone. Androgen replacement at a dose that maintained serum testosterone at approximately half the baseline levels was sufficient to sustain normal sexual function. There was no evidence that the experimental manipulation of testosterone had demonstrable effects on aggression or irritability.

Another approach to explore threshold hormonal effects on male sexuality is the induction of supraphysiological hormonal levels by testosterone administration. Anderson and colleagues[12] found that weekly testosterone injections over a 2-month period enhanced sexual interest and arousability in normal men without changes in overt behavior and without significant effects on mood and aggression. The results of the above two studies are consistent with previous observations on hypogonadal patients on the importance of testosterone for maintaining sexual desire but they differ in their conclusions regarding the significance of circulating levels for sexual function. While the experimental induction of a hypogonadal state suggests that the threshold for impairing sexual function may be below the normal range of circulating testosterone, the results of androgen administration indicate that increases of sexual interest and arousability remain possible above the normal physiological range for this hormone. It should be noted that both studies involve acute experimental manipulations and the results may not be generalizable to long-term hormonal effects or to the role of testosterone in sexually dysfunctional individuals.

ANDROGENS AND MALE SEXUAL DYSFUNCTION

Studies of pituitary–gonadal function in physically healthy men with sexual dysfunction have furnished inconsistent evidence. Although lower circulating testosterone has been reported by some investigators, most have been unable to detect hormonal differences between impotent men free from organic problems and non-dysfunctional subjects[13]. Despite research with hypogonadal patients which showed that the physiologic processes that subserve sexual drive are more sensitive to androgen deprivation than those that mediate erectile capacity, few studies have explored endocrinologically the possible coexistence of low sexual desire in men with erectile difficulties. Furthermore, there is virtually no hormonal information on men with the primary diagnosis of hypoactive sexual desire disorder (HSDD). We have found[14] that men with HSDD had significantly lower plasma testosterone measured hourly during the night than controls and that there was a relation between the decrease in testosterone and the impairment of sexual drive. In view of the suggestion that NPT may reflect central androgen effects, it was of note that the NPT parameters of the patients with HSDD and secondary erectile difficulties were significantly lower than those of the non-dysfunctional men. Precise clinical characterization and the combined application of hormonal and psychophysiological methods of evaluation may be required to identify biologically distinct sexually dysfunctional subgroups of patients.

ANDROGENS AND AGING SEXUALITY

There is consistent evidence in the literature of overall age-related decreases in sexual behavior with wide variations in individual rates of decline. Pfeiffer and collaborators[15] documented in two extensive multidisciplinary studies conducted at Duke University, a gradual decline in sexual interest and intercourse frequency with advancing years. Intensity of sexual interest decreased to a lesser extent than sexual activity, with some men in their 80s still retaining a moderate degree of sexual desire. Martin[16] obtained sexual information from a cross-sectional assessment of 628 men aged 25–85 years who participated in the Baltimore Longitudinal Study of Aging. As in the Duke studies he observed an age-related decrease in sexual behavior as well as marked individual variability within age groups. Erectile capacity was an important factor associated with frequency of sexual activity during the year prior to the interview. The percentage of men who reported erectile problems increased from 7% at ages 20–30 to 57% at ages 70–79. More recently McKinlay and Feldman[17] obtained systematic sexual data from a large sample of volunteers aged 40–70 who participated in an epidemiological study in randomly selected communities in the Boston area. There was a significant age-related decrease in erectile function, a decline in frequency of intercourse and ejaculation and an increase in ejaculatory difficulties. Sexual desire also decreased with age, a change that closely paralleled the decline in sexual activity.

There is also considerable, although not entirely consistent, evidence of a decline in gonadal function as age progresses[18,19]. These behavioral and hormonal data have led to the hypothesis that androgen deficiency is responsible for the decrease in sexual desire and activity in older men. Remarkably few studies[20,21] have systematically assessed the role of hormones on age-related differences in sexual behavior. Many of these data have been limited by the confounding effects of medical illness and drugs, the small number of hormonal determinations, the lack of objective measures of erectile capacity or the narrow range of sexual questions without validation by the sexual partner.

We have reported on an investigation[22] that gathered, within a cross-sectional framework, behavioral, hormonal and psychophysiological data on 77 healthy men aged 45–74 living in stable sexual relationships. The couples were selected from a pool of volunteers who responded to media announcements requesting participation in an investigation of factors that contribute to health, well-being and marital satisfaction in older men. All subjects and their spouses underwent extensive interviews to gather psychosexual, medical and marital information and answer a battery of psychological tests. After completion of a comprehensive medical and urological evaluation and penile vascular assessments to rule out vascular problems, all men were studied in the Sleep Laboratory for 4 nights. Electroencephalograms, eye movements, muscle tone and penile tumescence were monitored continuously through all the nights. During the 4th night blood samples were obtained at 20-min intervals, *via* a catheter from outside the subjects' room, for hormonal determinations. Plasma testosterone, bioavailable testosterone (bT), estradiol, luteinizing hormone (LH) and prolactin were measured in duplicate by means of established radioimmunoassay procedures. The following is a summary of the main findings.

Relationship between age and sexual behavior

The relation between age and the sexual dimensions assessed in the psycho-sexual interview were evaluated by correlation analysis of the total sample and with non-parametric comparisons of values in the 45–54, 55–64 and 65–74 age groups[22].

The majority of the sexual desire, sexual arousal and sexual activity dimensions demonstrated a highly significant negative correlation with age. Non-parametric multiple comparisons revealed that the differences in sexual behavior were primarily between the 45–64 and the 65–74 age groups. There was a significant age-related increase in reported prevalence of erectile problems. Prospective behavioral assessment conducted daily for 1 month validated the retrospective information obtained during the psychosexual interview.

Relation between age, sleep and nocturnal penile tumescence measures

There was a significant decrease in sleep efficiency with aging but no difficulties in REM sleep. Highly significant age-related decreases in frequency, duration and degree of nocturnal erectile episodes were observed. The decrease in duration of tumescence remained significant after adjustment for the decrease in total sleep time associated with aging. Ninety per cent of the subjects less than 65 years of age had at least one episode of maximum tumescence during the 4 study nights; in contrast, in only three of the subjects who were 65 or older was there clear evidence of penile rigidity during sleep.

Relation between age, hormones and sexual behavior

As may be seen in Table 1 age was negatively correlated with bT and with the ratio of bT over LH, it was positively correlated with plasma LH and was not related to total testosterone, estradiol and prolactin[23,24]. Bioavailable testosterone and its ratio over LH showed a close association with several desire and arousal dimensions and with reported sleep erections but they were not related to sexual activity. Total testosterone, estradiol and prolactin demonstrated few or no behavioral relationships. The hormonal–behavioral relationships were evaluated further by comparing the bT levels in two groups of subjects divided on the basis of the frequency or degree of behavioral measures. Men who reported in the interview desiring sexual experiences with their wives or masturbating with a greater frequency than once per week had higher bT levels than men with lower frequency rates. Similarly the bT concentration was higher in men who reported having sleep-related erections more than once per week than in men with less frequent nocturnal erections. All the hormonal–behavioral differences between groups lost their significance following adjustment, by analysis of covariance, for the effect of age on both endocrine and sexual behavior measures. Prospective as well as retrospective data on the prevalence of sexual dysfunction were unrelated to all hormonal variables.

Relation between age, hormones and nocturnal penile tumescence

Bancroft[25] hypothesized that the threshold required for the behavioral effects of testosterone increases with age; although older men may have plasma concentrations within the 'normal' range, those levels may no longer be sufficient for adequate sexual function. The observation that sleep erections are androgen-sensitive and our findings that they are impaired in healthy older men is consistent with this hypothesis. However, the notion that older men may have a relative decrease in testosterone action that contributes to the decline in nocturnal erectile function has yet to be subjected to empirical evaluation.

We examined[26] the relationship between pituitary and gonadal hormone levels and NPT measures in the above described study in order to test the hypothesis that the association between bT and

Table 1 Correlations between plasma hormones, age and sexual behavior; *p*-values in parentheses

Psychosexual variable	Total testosterone	Estradiol	Bioavailable testosterone	LH	Bioavailable testosterone/LH	Prolactin	Age
Retrospective assessment							
Desire							
frequency of sexual thoughts			0.43 (0.008)		0.36 (0.03)		−0.50 (0.001)
frequency of desire for sex			0.52 (0.001)		0.45 (0.006)		−0.52 (0.001)
Arousal							
ease of becoming aroused			0.26 (0.04)		0.24 (0.05)		−0.45 (0.0001)
degree of masturbation erections					0.33 (0.03)		−0.38 (0.01)
degree of coital erections			0.34 (0.01)	−0.33 (0.01)	0.35 (0.01)		−0.59 (0.0001)
frequency of sleep erections			0.30 (0.01)	−0.24 (0.05)	0.33 (0.006)		−0.36 (0.002)
Activity							
frequency of coitus							−0.40 (0.0004)
frequency of masturbation	−0.34 (0.0005)						−0.34 (0.002)
no. of orgasms per month							−0.51 (0.0001)

(continued)

(*continued*)

Prospective assessment						
frequency of sexual thoughts	0.31 (0.02)					−0.14 (0.28)
frequency of sleep erections			0.28 (0.04)	−0.33 (0.01)	0.35 (0.01)	−0.44 (0.0004)
frequency of coitus						−0.42 (0.001)
frequency of masturbation						−0.22 (0.09)
Age	−0.09 (0.45)	−0.20 (0.10)	−0.60 (0.0001)	0.32 (0.008)	−0.51 (0.0001)	0.04 (0.76)

Behavioral–hormonal correlations indicated only when significant. Reproduced in modified form from ref. 23

NPT measures will be more evident in older than younger men who, themselves, may have bT levels well above the central threshold of NPT activation. We found that bT correlated significantly with several NPT measures while total testosterone, estradiol, LH and prolactin were mostly unrelated to NPT (Table 2). All bT–NPT correlations lost their significance after adjustment by multiple regression analysis for the effect of age on both bT and NPT in the total subject sample. Analysis of bT–NPT relations by age groups revealed a significant positive association between bT and duration of NPT in men aged 55–64 years, but no statistical relations in younger and older age groups. We interpreted this finding as reflecting a change in the central threshold of hormonal activation of NPT as age progresses. Circulating androgens may be well above the threshold of activation on NPT responses in younger men, fall within the elevated threshold range of middle-aged individuals, or become no longer sufficient to sustain adequate NPT function in older men.

To summarize, healthy men in stable relations, between 45 and 74 years of age, demonstrated age-related decline in sexual desire, sexual arousal and sexual activity. There was a highly significant decrease in bT and a compensatory increase in LH consistent with the notion that aging is associated with a reduction of gonadal activity. This age-related decline cannot be explained by the effect of medical illness or drugs as postulated by several authors in view of the rigorous inclusion/exclusion criteria and the extensive medical screening procedures. Analysis of the relation between circulating hormones and behavior provided information consistent with the hypothesis that androgen deficiency may contribute to the age-related decline in male sexuality. Sexual desire and reported sleep erections were the behavioral domains more closely related to bT and bT/LH. In contrast, frequency of coitus, which is presumably influenced by the partner's own receptivity, did not show significant hormonal associations. The age-related effect of bT was, however, a more important determinant of reported variations of sexual behavior and sleep erections than were the effects of bT independent of age. There was no evidence that hormonal factors contributed to erectile dysfunction in healthy aging subjects. The notion, suggested by the present findings, that older men may have a relative decrease in central testosterone action that contributes to the decline in both sexual desire and nocturnal erectile function, deserves further empirical evaluation.

Table 2 Correlations between plasma hormones, age and nocturnal penile tumescence (NPT); p values in parentheses

NPT parameters	Total testosterone	Estradiol	Bioavailable testosterone	LH	Prolactin	Age
Frequency						
Total episodes			0.31 (0.01)			−0.35 (0.002)
Episodes during REM			0.34 (0.005)			−0.38 (0.0008)
Episodes with maximum tumescence			0.30 (0.02)			−0.45 (0.0001)
Duration						
Total tumescence			0.37 (0.002)			−0.40 (0.0003)
Simultaneous REM and tumescence			0.39 (0.001)	−0.24 (0.05)		−0.44 (0.0001)
Maximum tumescence			0.31 (0.01)			−0.43 (0.0002)
Tumescence/sleep time (%)			0.32 (0.008)			−0.35 (0.002)
Degree						
Mean increase in circumference			0.24 (0.055)	−0.33 (0.006)		−0.36 (0.001)
Maximum increase in circumference				−0.34 (0.006)		−0.36 (0.001)
Age	−0.09 ns	−0.20 ns	−0.60 (0.0001)	0.32 (0.008)	0.04 ns	

Behavioral–hormonal correlations indicated only when significant; ns = $p > 0.05$; reproduced in modified form from ref. 26

ANDROGEN SUPPLEMENTATION AND SEXUAL FUNCTION

It has been hypothesized that a relative androgen deficiency may contribute to the decline of sexual function in aging but evidence in support of this notion is lacking. Placebo-controlled studies of androgen administration to sexually dysfunctional men are few and there are, to our knowledge, none that have examined androgen effects on the sexual behavior of healthy older men. O'Carroll and Bancroft[27] carried out a double-blind, cross-over comparison of intramuscular injections of testosterone and placebo in two groups of relatively young eugonadal men diagnosed as having either low sexual interest or erectile dysfunction. Androgen injections resulted in a substantial elevation of circulating testosterone and a modest increase in sexual interest in the low desire group but no effect in erectile capacity in either group. Carani and co-workers[28] found that androgen treatment only enhanced the sexual function of impotent men with borderline levels of free testosterone but not in men with free testosterone within the normal range. We have recently completed[29] a double-blind, placebo-controlled, cross-over study to assess the effect of biweekly injections of testosterone enanthate (200 mg) over a 6-week period on sexual behavior and mood in a small group of healthy eugonadal men aged 46–67 (median 60 years) with erectile dysfunction. Testosterone administration resulted in significant increases in frequency of masturbation, sexual activity with partner and early morning erections but had no effects on reported erectile rigidity. To the extent that the increase in sexual activity reflected sexual interest, these findings are consistent with information from hypogonadal men suggesting that the primary effect of testosterone is on sexual desire and sleep-related erection rather than on erectile responses to external stimuli. There was no evidence that exogenous testosterone had significant effects on the subjects' affective state or on mood. It should be mentioned, however, that the drug regimen did not maintain supraphysiologic testosterone levels throughout the hormonal phase of the investigation. Further long-term, controlled, interdisciplinary research is required to address this largely unexplored aspect of behavioral endocrinology.

ACKNOWLEDGEMENT

Research conducted in our laboratories was supported by National Institute in Aging Grant 06895.

REFERENCES

1. Skakkeback, N. E., Bancroft, J., Davidson, D. W. and Warner, P. (1981). Androgen replacement with oral testosterone undecanoate in hypogonadal men: a double-blind controlled study. *Clin. Endocrinol.*, **14**, 49–61

2. Salmimies, P., Kockott, G., Pirke, K. W., Vogt, H. J. and Schill, W. B. (1982). Effects of testosterone replacement on sexual behavior in hypogonadal men. *Arch. Sex. Behav.*, **11**, 345–53

3. Kwan, M., Greenleaf, W. J., Mann, J., Crapo, L. and Davidson, J. (1983). The nature of androgen action on male sexuality: a combined laboratory/self report study on hypogonadal men. *J. Clin. Endocrinol. Metab.*, **57**, 557–62

4. O'Carroll, R., Shapiro, C. and Bancroft, J. (1985). Androgens, behavior and nocturnal erections in hypogonadal men: the effect of varying the replacement dose. *Clin. Endocrinol.*, **23**, 527–37

5. Davidson, J. M., Camargo, C. and Smith, E. R. (1979). Effects of androgen on sexual behavior in hypogonadal men. *J. Clin. Endocrinol. Metab.*, **48**, 955–8

6. Davidson, J. M., Kwan, M. and Greenleaf, W. J. (1982). Hormonal replacement and sexuality. *Clin. Endocrinol. Metab.*, **11**, 599–623

7. Bancroft, J. and Wu, F. C. W. (1983). Changes in erectile responsiveness during androgen replacement therapy. *Arch. Sex. Behav.*, **12**, 59–66

8. Rubin, H. B., Henson, D. E., Falvo, R. E. and High, R. W. (1979). The relationship between men's endogenous levels of testosterone and their penile responses to erotic stimuli. *Behav. Res. Ther.*, **17**, 305–12

9. Lange, J. D., Brown, W. A., Wincze, J. and Zwick, W. (1980). Serum testosterone concentration and penile tumescence changes in men. *Horm. Behav.*, **14**, 267–70

10. Carani, C., Scutery, A., Marrama, P. and Bancroft, J. (1990). The effects of testosterone administration and visual erotic stimuli on nocturnal penile tumescence in normal young men. *Horm. Behav.*, **24**, 435–41

11. Bagatell, C. J., Heiman, J. R., Rivier, J. E. and Bremner, W. J. (1994). Effects of endogenous testosterone and estradiol on sexual behavior in normal young men. *J. Clin. Endocrinol. Metab.*, **78**, 711–16

12. Anderson, R. A., Bancroft, J. and Wu, F. C. W. (1992). The effects of exogenous testosterone on sexuality and mood of normal men. *J. Clin. Endocrinol. Metab.*, **75**, 1503–7

13. Davidson, J. and Rosen, R. C. (1992). Hormonal determinants of erectile function. In Rosen, R. C. and Leiblum, S. R. (eds.) *Erectile Disorders, Assessment and Treatment*, pp. 72–95. (New York and London: The Guilford Press)

14. Schiavi, R. C., Schreiner-Engel, P., White, D. and Mandeli, J. (1988). Pituitary–gonadal function during sleep in men with hypoactive sexual desire and in normal controls. *Psychosom. Med.*, **50**, 304–18

15. Pfeiffer, E., Verwoerdt, A. and Wang, H. S. (1968). Sexual behavior in aged men and women. *Arch. Gen. Psychiatry*, **19**, 753–8

16. Martin, C. E. (1975). Marital and sexual factors in relation to age, disease and longevity. In Wirdt, R. D., Winokur, G. and Ruff, M. (eds.) *Life History Research in Psychopathology*, pp. 326–47, Vol. 4. (Minneapolis: University of Minnesota Press)

17. McKinlay, J. B. and Feldman, H. A. (1994). Age-related variation in sexual activity and interest in normal men: results of the Massachusetts male aging study. In Rossi, A. S. (ed.) *Sexuality Across the Life Course*, pp. 261–86. (Chicago and London: The University of Chicago Press)

18. Vermeulen, A. (1991). Clinical Review 24. Androgens in the aging men. *J. Clin. Endocrinol. Metab.*, **73**, 221–4

19. Vermeulen, A. and Kaufman, J. M. (1992). Editorial: role of the hypothalamo–pituitary function in the hypoandrogenism of healthy aging. *J. Clin. Endocrinol. Metab.*, **74**, 1226A–C

20. Tsitouras, P. D., Martin, C. E. and Harman, S. M. (1982). Relation of serum testosterone to sexual activity in healthy elderly men. *J. Gerontol.*, **37**, 288–93

21. Davidson, J. M., Chen, J. J., Crapo, L., Gray, G. D., Greenleaf, W. J. and Catania, J. A. (1983). Hormonal changes and sexual function in aging men. *J. Clin. Endocrinol. Metab.*, **57**, 71–7

22. Schiavi, R. C., Schreiner-Engel, P., Mandeli, J., Schanzer, H. and Cohen, E. (1990). Healthy aging and male sexual function. *Am. J. Psychiatry*, **147**, 766–71

23. Schiavi, R. C., Schreiner-Engel, P., White, D. and Mandeli, J. (1991). The relationship between pituitary–gonadal function and sexual behavior in healthy aging men. *Psychosom. Med.*, **53**, 363–74

24. Schiavi, R. C., White, D. and Mandeli, J. (1993). Pituitary–gonadal function during sleep in healthy aging men. *Psychoneuroendocrinology*, **17**, 599–609

25. Bancroft, J. (1984). Androgens, sexuality and the ageing male. In Labrie, F. and Proulx, L. (eds.) *Endocrinology*, pp. 913–16. (Amsterdam: Elsevier)

26. Schiavi, R. C., White, D., Mandeli, J. and Schreiner-Engel, P. (1993). Hormones and nocturnal penile tumescence in healthy aging men. *Arch. Sex. Behav.*, **22**, 207–15

27. O'Carroll, R. and Bancroft, J. (1984). Testosterone therapy for low sexual interest and erectile dysfunction in men: a controlled study. *Br. J. Psychiatry*, **145**, 146–51

28. Carani, C., Zini, D., Baldini, A., Della Casa, L., Ghizzani, A. and Marrama, P. (1990). Effects of androgen treatment in impotent men with normal and low levels of free testosterone. *Arch. Sex. Behav.*, **19**, 223–34

29. Schiavi, R. C., White, D., Mandeli, J. and Levine, A. C. (1996). Effect of testosterone administration on sexual behavior and mood in men with erectile dysfunction. *Arch. Sex. Behav.* (in press)

DISCUSSION

Schröder: I am somewhat surprised that you did not find an increase of plasma testosterone with the testosterone enanthate in your supplementation study[1]. Wallace *et al.*[2] administered the same dose of 200 mg of testosterone enanthate to young men and found a highly significant increase of plasma testosterone.

Schiavi: We have chosen 200 mg every two weeks, because of data in hypogonadal men that showed this dose results in sustaining levels of testosterone after two weeks. However, in the men we studied, whose median age was 60, it emerged not to be the case, even though there were significant decreases in LH.

Tenover: I think it is important to note that the men in your study were not hypogonadal. Furthermore, you measured testosterone at the end of each two week treatment period, whereas you need to know the area under the curve for total testosterone over the entire two weeks on therapy *versus* two weeks on placebo. I suspect that if you had measured testosterone in the first three or four days after administration, it would have been higher. Obviously, one does not expect much of response when testosterone levels do not change much. More generally, we have to look whether total levels of testosterone really changed over time when we get negative data in terms of response to administration. The change of androgen levels over time is crucial when evaluating testosterone administration effects on mood, muscle, bone, etc. We must take this aspect into account when interpreting the literature. At least, you need some sort of change of plasma levels, otherwise you do not know if the data are negative because all you have done is to replace endogenous testosterone with exogenous testosterone.

Schiavi: Indeed, it is very important to clarify that those individuals who were accepted in our study were not selected on the basis of having some hypogonadal changes which explains that testosterone remained in the normal range in our study. The marked decrease in circulating LH does suggest that testosterone administration had a central hormonal effect. It is quite likely that plasma testosterone levels had returned to baseline two weeks after hormonal administration.

Morales: Could you speculate more about the correlation between nocturnal erections and waking erections? The feeling is that the two are under different neuro-hormonal controls. Repeatedly, evidence has been presented that agonadal individuals are still able to have sexual erections although they lose their nocturnal erections very quickly.

Schiavi: Indeed, we cannot assume that the psychobiological mechanisms that mediate nocturnal erections are the same as the mechanisms that mediate erotical induced erections. However, this is based on an assumption and we do not have any evidence to substantiate it. Therefore, there may or may not be different hormonal correlates of both events. In our study, we investigated why some of the men who had significantly depressed nocturnal erections were able to function adequately with a partner. These men reported that they were able to compensate for what seems an organical impairment, by increasing their degree of sexual activity, use of erotica and enhancing the erotic tension of the relationship. This points in the direction of an interplay of organic deteriorating factors that may be age-dependent and compensating psychological factors that enhance erectile capacity.

Orwoll: I have a comment referring to the previous question of the appropriate dose for treating elderly men with the intent to actually increase testosterone levels. We have just finished a study with 40 older men with baseline testosterone levels below the age-related mean but with normal LH concentrations. We treated those men with either placebo or the scrotal testosterone patch, which is known to increase serum testosterone concentrations in hypogonadal men. In the men investigated there was no change in serum testosterone concentrations at all, and in fact, by measuring repeatedly for a 24 hour period, we were able to demonstrate that all we did was to abolish the circadian rhythm in testosterone while suppressing their LH concentrations. We actually replaced their endogenous testosterone production with the exogenous testosterone administered, bringing us nowhere in the end. Incidentally, there were no changes in sexual function in those men. So choosing a testosterone dose in the older man may be very different from choosing a testosterone dose in the hypogonadal man, and we need studies clearly designed to establish those dose–response relationships.

Schröder: Did you measure prostate specific antigen (PSA) in the study you just described?

Orwoll: We measured PSA in these 40 men and in an additional 60 men we also studied, and there were no changes in either group, despite the fact that there were clearly quite significant elevations in dihydrotestosterone levels. Neither did we see any changes in urinary flow or prostate size.

Bain: Dr Schiavi, did you analyse the response to supplementation separately for those men in the study who had lower levels of testosterone at baseline? Did you observe different responses in this group as compared with the total study sample?

Schiavi: Out of the twelve men we studied, there were four or five who had levels that were borderline normal. We assessed the effect of testosterone supplementation in those men, and there were no differences from the total group.

Thijssen: In the cross-sectional study you described[3] you selected 77 out of 300 individuals, which is about one out of four. What about the excluded men? Is it possible to extrapolate the findings in the 77 selected men to the whole population?

Schiavi: We cannot generalize the results of this study to the population at large. Nevertheless, it is important to the extent that it allows assessing behavior-endocrine relationships in healthy individuals. Reasons for excluding the other men included the partner's unwillingness to participate and medical illnesses, or medication use that would have confounded the results of our evaluation.

Oddens: It has been suggested that androgens may influence mood and well-being. I wondered whether effects of androgens on sexual desire could be mediated by their effects on well-being. In other words, is it sexual desire that changes due to a lack of androgens, or is it primarily well-being that changes, which may then in turn affect sexual desire?

Schiavi: The question has not yet been empirically answered to everyone's satisfaction. Most of the studies available in relation to male sexual function focus on sexual desire *per se* but they have not equally assessed mood-related changes, and to what extent the latter may affect sexual desire.

Christiansen: In our research on androgens and mood in healthy males, we also evaluated sexual desire. We found out that with higher testosterone levels (though still within the normal physiological range) the men reported a higher sexual desire too.

Morales: From our research in men in their mid-fifties who were impotent and received androgens supplemental therapy, we observed that the men who responded to treatment had a parallel improvement not only in mood, but also in the frequency of sexual thoughts, the excitement derived from those thoughts, and the frequency of vaginal penetrations.

REFERENCES

1. Schiavi, R. C., White, D., Mandeli, J. and Levine, A. C. (1996). Effect of testostereone administration on sexual behavior and mood in men with erectile dysfunction. *Arch. Sex Behav.* (in review)
2. Wallace, E. M., Pye, S. D., Wild, S. R., Wu, F. C. W. (1993). Prostate-specific antigen and prostate gland size in men receiving exogenous testosterone for male contraception. *Int. J. Androl.,* **16**, 35–40
3. Schiavi, R. C., Schreiner-Engel, P., Mandeli, J., Schanzer, H. and Cohen, E. (1990). Healthy aging and male sexual function. *Am. J. Psychiatry,* **147**, 766–71

8

Flushes in men and women: circulatory and etiological aspects

J. Ginsburg

CLINICAL FEATURES

The hot flush is the commonest symptom of the menopause occurring in at least 70% of women[1] and sufficiently severe in over half of those afflicted for them to request treatment from their doctors. Although hot flushes are in general most frequent and troublesome in the first few months after the final menstrual period, with the onset of the menopause, they can be equally severe in the perimenopausal phase and can also recur or present for the first time in women in their 60s and 70s. There is no uniformity about the duration of time over which flushes continue. They may be present for only a few weeks or months but can also persist for years, becoming more severe with time, even in women who initially had slight symptoms when they entered on the menopause.

The classical clinical description of the phenomenon is a sudden sensation of increased heat, manifest to the observer by reddening of the face and also frequently of the neck but often experienced by the woman in the upper part of the body as well and sometimes spreading 'all over'. The facial reddening and sensations of heat are generally transient but sweating is frequently associated and if this is severe, the woman may have symptoms for more than 2–3 min – sometimes even as long as 15 min. After the sensations of heat have subsided, she may feel cold and even shiver. There is often a prodromal phase before the woman notices the sensation of increased heat and which may indicate to her that the flush is about to occur. This sometimes takes the form of an unpleasant ill-defined sensation in the chest. The frequency with which flushes occur varies considerably even in the same woman. There may be only isolated episodes during the day, or occasionally during the week in some women, whilst in others, the flushes occur frequently – at least once an hour and sometimes several times an hour. Some women are disturbed frequently at night with flushes. In general, however, even if women have no nocturnal flushes, they are much hotter than their non-flushing peers and wake up with a desire to throw off their bed covers. Hot drinks, stress and increased ambient temperature may precipitate flushes in susceptible individuals.

Geographical incidence and cultural differences

The initial epidemiological and social studies of menopausal flushing were performed in industrialized communities, in Europe and North America. In Western cultures, the prevalence of hot flushes was reported to be at its height in the first few years after the onset of the menopause, occurring in between 50 and 93% of women but with decreasing frequency over time. The most severe manifestation is generally held to occur in women after bilateral ovariectomy, a phenomenon attributed to the sudden 'cut off' in ovarian estrogen production as the result of surgery. There seems to be some correlation with social class, the incidence of hot flushes being reported as less in the professional, educated woman than in the factory worker although it should be noted that the educated professional woman, and particularly those from North America, may be the most articulate and active in search of treatment for hot flushes.

Studies of the nature and frequency of hot flushes in other cultures – in the Indian subcontinent, native American and Japanese women for example – have revealed wide variation both in prevalence and severity of hot flushes. Thus Japanese women apparently report very little in the line of hot flushes[2], or at least they did in the past. But is the phenomenon really *infrequent* in Japan or is it the perception of flushes by the woman which differs or is it her reluctance to admit to such symptoms when she has become menopausal? Or is there perhaps some other explanation – for example a dietary factor such as a phyto-estrogen – which may prevent the full flowering of the episodes as we know them in the West? The complaints and prevalence of hot flushes in different cultures may also relate to the attitude of particular societies (and also groups within them) to the menopausal woman, i.e. whether this is a privileged state as in the Yucatan, where the menopausal woman has enhanced status in society, or whether it is a stage of her life, as in some Western communities, when she is no longer sought after or admired, in which case her status symbol and role become downgraded. In the former, the elderly Mayan women do not apparently complain of flushes[3], whereas Western women do, vociferously. Then there is the 'deserted nest syndrome' of the woman/man whose children are pursuing their own career away from home or whose wife/husband may have left for a younger version! The problem of isolating these different factors makes it difficult to provide precise statements of the prevalence of hot flushes and of their duration and intensity in women of different geographical regions and cultures.

CIRCULATORY CHANGES DURING THE MENOPAUSAL HOT FLUSH

Scientific studies of the hot flush are a relatively recent phenomenon. Credit for what is probably the first attempt to analyze what happens during the flush should be given to Molnar who made a detailed study of body and skin temperature responses to spontaneous flushes ('flashes' as in the American idiom) and to an evoked flush 20 years ago[4] in a naked woman (his wife!) (Figure 1). He showed that after the onset of the flush there is a rise in skin temperature recorded in thermocouples placed on the finger, toe and forehead. Temperature in the extremities rapidly reaches a peak, sustained for a few minutes and then falls asymptotically towards the baseline control temperature over a matter of 20 min in the case of the finger and 50 min at the toe. Forehead temperature was still elevated at an hour and indeed fell during a subsequent evoked flush. By contrast there was no rise in body temperature as indicated by thermocouples placed in the rectum or on the tympanum. In fact the reverse is the case, a progressive fall in body temperature over time which may well reflect the fact that the woman was naked. There was, however, a more marked fall in the thermocouple placed on the tympanum during both spontaneous and evoked flushes. The subsequent fall in temperature of the forehead with the evoked flush may be explained by the fact that sweating, which can be severe and can continue after the sensations of the hot flush have subsided, results in evaporative cooling causing the temperature of the wet skin to drop, particularly on the forehead where sweating is generally profuse. Molnar's study also showed a disturbance of the baseline electrocardiographic record, a change confirmed by Sturdee and co-workers[5].

Subsequently other workers were stimulated to investigate the thermal accompaniments of the hot flush, skin temperature measurements being

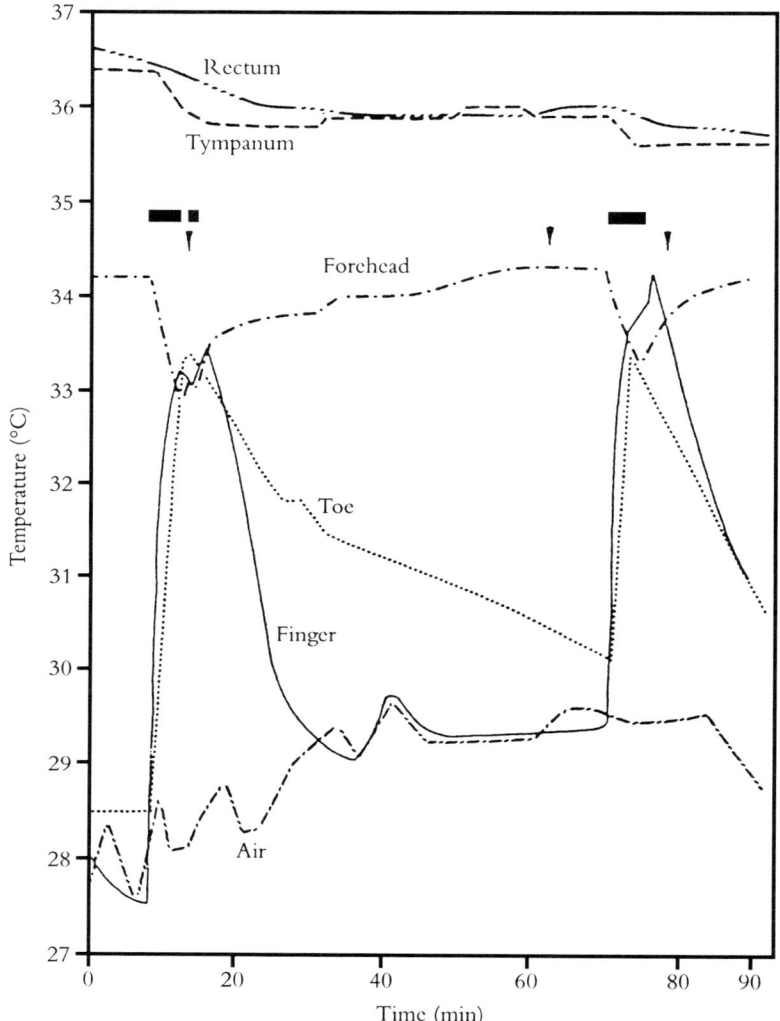

Figure 1 Peripheral skin temperature and tympanum and rectal temperature recorded in a menopausal woman during 90 min in relation to menopausal hot flushes. Reproduced with permission from reference 4

claimed to provide an objective record of the circulatory events occurring during the hot flush and taken as an accurate evaluation of drug effects[6–8]. The assumption that skin temperature measurements reflect what happens from the circulatory point of view during the hot flush is not however justified. Measurement of skin temperature *per se* is of very limited value and inference from changes in skin temperature should be made cautiously. It must be emphasized that all we can infer from cutaneous temperature is the difference between heat loss to the environment by convection, evaporation, radiation, conduction and that which is conveyed to the skin by the incoming arterial blood. Hence, if a rise in skin temperature is used as an index of the rate of blood flow, it is essential that the amount of heat that is lost to the environment during this time does not change. Since, however, sweating is frequent during the hot flush and this, as stated above, increases evaporation from the skin and thereby changes skin temperature, this assumption is not justified. Furthermore, even in the absence of sweating, an increased blood flow to a local area of skin will of course increase the amount of heat lost by radiation.

There are also technical problems because of the nature of the recording device used and the fact that in any case temperature of the skin reaches a maximum long before skin blood flow has reached its peak so that no further changes in surface temperature can then be recorded when cutaneous flow increases. Also, with rapid fluctuations in flow, particularly at high rates of cutaneous blood flow, skin temperature cannot respond to these rapid changes. It is therefore insensitive and unreliable as an indicator of changes in the local circulation. Workers in this field have also ignored the fact that changes in skin temperature provide no information regarding any circulatory changes in tissues deep to the skin.

In the light of the above it is therefore not surprising that the correlation between hot flushes and skin temperature is relatively poor. But despite the limitations of the method skin temperature measurements have provided important information in that they have served to emphasize the profound thermoregulatory disturbance present in flushing women. This is particularly marked during the night when there are frequent temperature spikes recorded even during sleep in flushing women and which are then abolished by estrogen treatment.

In order to evaluate what precisely happens from the circulatory point of view during the hot flush, it is therefore essential to use a method which is capable of measuring rapid circulatory changes, such as venous occlusion plethysmography. Using this method, we have shown[9] that there is a characteristic pattern of circulatory change during the flush itself – a rapid increase in hand blood flow observed with the onset of symptoms and then shortly after a lesser but still significant rise in both forearm flow and pulse rate (Figure 2). Hand flow falls slightly after reaching a peak which may be 10 times that of resting flow and then remains elevated above resting levels for some 5 min before falling to baseline values about 8–10 min after the sensation of heat has subsided. Meanwhile forearm flow and pulse rate have, however, fallen to control values at a time when hand flow is still elevated. Interestingly enough there is no change in blood pressure – either systolic or diastolic – during the period of the flush itself or subsequently. It should also be emphasized that despite the symptoms experienced by women in the prodromal phase, just before the occurrence of sensations of heat and visible reddening of the face, we were unable, in those women who had a clear-cut symptomatic prodromal phase in the course of their laboratory studies, to demonstrate any circulatory change – either in peripheral blood flow, pulse rate or blood pressure *before* the actual onset of the flush itself.

The onset of the vascular changes we measured was apparently abrupt and the responses were clear cut and similar in all those women in whom we were able to obtain accurate circulatory recordings. We should however point out that from the technical point of view such measurements are very difficult since women often feel uncomfortable with the onset of the flush and may start to move around thereby disturbing the recording set-up and often making measurements virtually impossible.

The pattern of circulatory change we observed is typical of the circulatory changes recorded in response to indirect heating, which would suggest activation of the reflex mechanisms for dissipating excess body heat. There is, however, no apparent 'trigger' factor, such as an increase in core temperature. Indeed, as explained above, there is a lowering of body temperature. Since there is no increase in body temperature, it would seem that 'false' information regarding a change in body temperature is given to the central sensors concerned with heat regulation. This fits with the clinical observation that women who flush may feel much hotter than their non-flushing peers even in the absence of any episodes of flushing as such. Women who flush frequently, often complain of the heat in a room which others find tolerable and insist on opening windows and doors so that the ambient temperature may drop to unacceptably cool levels for others. Flushing women often wear very much thinner clothes when doing their housework since they feel so warm. Marital discord can result from the menopausal woman demanding much thinner bed clothes than her husband in the height of winter.

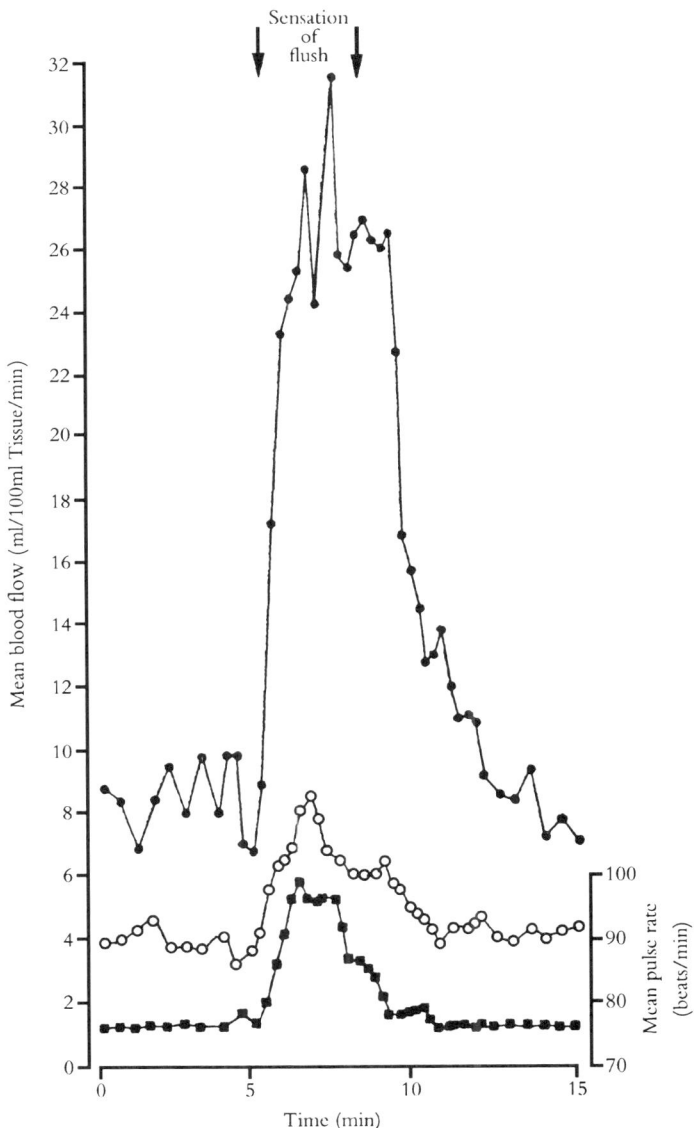

Figure 2 Mean hand (solid circles) and forearm (open circles) blood flow and pulse rate (solid squares) in six women before, during and after a menopausal hot flush. Reproduced with permission from reference 9

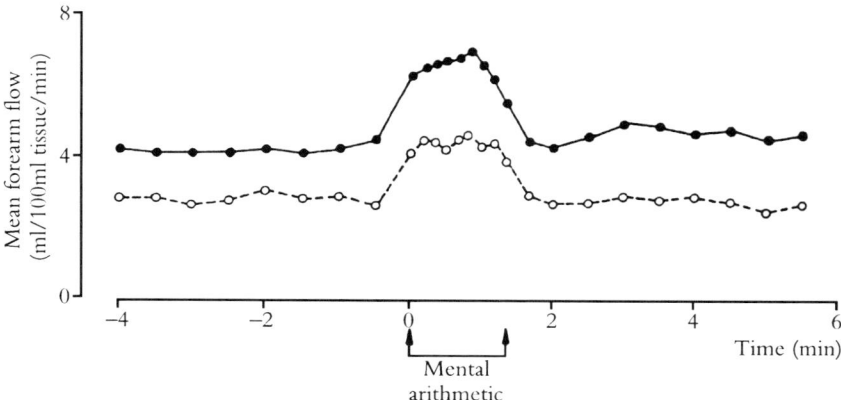

Figure 3 Mean forearm blood flow (ml/100 ml tissue/min) before, during and after stressful mental arithmetic in 15 menopausal women who had hot flushes (solid circles) and 13 who did not (open circles). Reproduced with permission from reference 10

Our further studies[10] have shown that circulatory changes are apparent in women who flush frequently, outside the acute events of the flush itself. Thus basal forearm flow in flushing women is significantly and persistently higher than in their non-flushing peers (Figure 3). More recently, we have found, in the course of studying the circulatory effects of drugs used to relieve menopausal symptoms such as hot flushes, that transdermal estradiol has a differential effect on the peripheral circulation dependent on whether women are flushing frequently or not[11]. In flushing women transdermal estradiol, given for 6–8 weeks, causes a lowering of the elevated forearm flow, but in those who do not flush frequently and who start with a lower basal forearm flow, transdermal estradiol induces a rise in forearm flow. There are no concomitant changes in blood pressure which might in part explain this differential effect. Estradiol does not apparently exert a direct effect on the peripheral vasculature in the woman, as indicated by the fact that the reactive hyperemia response in the forearm is unaffected by the administration of estradiol in either flushing or non-flushing women.

HORMONAL CHANGES

Innumerable attempts have been made to link an hormonal change, either with the initiation of the flush or the circulatory events characteristic of the phenomenon, but with little success. Although estrogen deprivation is thought to provide the essential causal background for the phenomenon, women with gonadal dysgenesis, whose basal levels of estrogen are normally extremely low, do not flush until *after* they have been treated with estrogens and this therapy is then stopped. It seems that 'priming' of the system with a period of adequate circulating estrogen is required before a reduction in estrogen concentration sets the scene for the appearance of classical hot flushes.

In postmenopausal women with severe debilitating flushes, blood estrone and estradiol levels, and in particular 'free' estradiol, were reported to be lower than in asymptomatic women of the same age[12–14]. There is however no correlation between circulating estrogen concentrations and the acute onset of individual flush episodes[15]. It should also be remembered that there are a number of conditions in which estrogen levels are low, as in anorexia nervosa, yet hot flushes are not a feature of anorexia nervosa or of other disturbances associated with low estrogens, aside from the menopause.

Although circulating cortisol, androstenedione and dehydroepian-drosterone (DHEA) were found to be increased in association with hot flushes, peak levels of these hormones were in general reached after the symptoms of the flush had ceased[15–17]. Similar changes were reported in respect of β-endorphin, β-lipotropin and adrenocorticotropic hormone (ACTH).

The report that a luteinizing hormone (LH) pulse is associated with individual flush episodes[18] was interpreted as denoting a causal relation between the two phenomena. Analysis of the reported data however shows that the peak of the LH pulse is displaced by at least 15 min or more from the onset of the flush itself. Furthermore, flushes occurred without a concomitant LH pulse and *vice versa*. Hot flushes have also been reported in women after hypophysectomy in whom there is no pulsatile LH release and also in women treated with a gonadotropin-releasing hormone (GnRH) agonist which suppresses pulsatile LH release, as well as in women with hypopituitarism. It would therefore seem that whatever the 'false' information fed to the hypothalamic centers regulating temperature, it is not invariably produced in a similar time sequence to that of LH pulses.

In the light of the foregoing, one might expect to find GnRH to be elevated just before the observed LH pulse and indeed this has been claimed to occur in women with hot flushes[19]. In view, however, of the poor temporal correlation between LH pulses and individual hot flushes it is most unlikely that GnRH is involved in any causal relation either in general in relation to the onset of hot flushes or in respect of individual flush episodes.

The possible involvement of catecholamines in menopausal flushing has also been investigated but with conflicting findings. Diametrically opposed results were observed with noradrenaline, some studies having shown no change in circulating levels of this amine[18,20] whilst another showed a decreased blood concentration[21]. Admittedly the circulation through the hand is essentially governed by release of constrictor tone and one would therefore expect noradrenaline levels to fall when hand flow increases but this does not imply cause and effect.

As far as adrenaline is concerned it is however difficult to relate a change in this catecholamine to the cardiovascular changes recorded during the hot flush since adrenaline causes a rise in systolic and a fall in diastolic pressure, reduces hand flow but causes a marked and continued increase in forearm flow.

Emotional blushing

For decades, the facial reddening associated with menopausal flushing and the fact that the phenomenon could readily be precipitated or intensified by emotional stress, was held to support the view of the majority of gynecologists that menopausal flushing was no more than a reactivation of the teenager's tendency to blush frequently. As long as this view is held, serious attempts to find the cause of menopausal flushing and thereby offer help to those severely afflicted with such problems will be limited. Similarly, we set out to delineate the cardiovascular changes during menopausal flushing and undertook a comparative study of emotional blushing.

Our study[22] was undertaken in volunteer colleagues and students who admitted to blushing frequently. Men as well as women were studied for this purpose. The subjects were told we were measuring basal levels of flow in those who blushed for comparison with our investigation of the comparative changes in the circulation as between women who flushed frequently and those who did not. Once control measurements had been made however we stressed the individual, either by staring fixedly at them, or asking embarrassing questions. A blush was readily provoked and circulatory measurements continued. The pattern of circulatory change during an emotional blush differed markedly from that seen during menopausal flushing. Thus hand flow did not rise as with the onset of flushing and indeed fell in the majority of incidences. Whilst forearm flow rose, this was to a lesser extent and for a shorter period than occurred with menopausal flushing and during the emotional blush there was a rise in mean blood pressure. Further analysis using strain gauge plethysmography in order to obtain an indication of the relative distribution of flow as between skin and muscle of the forearm, showed further that, whilst during the menopausal flush the main change was in the skin of the forearm, the change during emotional blushing would seem to be in the arterioles of skeletal muscle. This of course would account for the rise in blood pressure in the latter and not in the former. The pattern of response observed during emotional blushing, by contrast to that during flushing, is that of a response to stress as exemplified by the cardiovascular changes associated with stressful mental arithmetic. A flush is thus not a blush and whilst emotional factors may precipitate menopausal flushing, the cardiovascular response is not that classically observed in response to stress but reflects the integrated reflex involved in dissipating excess body heat.

FLUSHES IN MEN

That flushing can also occur in the male was first recognized in men who had experienced acute withdrawal of testosterone as after bilateral orchiectomy[23]. Similarly, down-regulation by GnRH agonists in men being treated for carcinomatosis of the prostate was also found to result in the occurrence of flushing episodes[24], during which men were drenched with sweat and behaved in all respects like flushing menopausal women. One would not expect flushing to be a frequent occurrence in the normal male as he ages; although testosterone falls with age in association with a parallel rise in gonadotropin concentrations, the lowered testosterone levels are not those of typical hypogonadism. Furthermore, the rate of testosterone decline in men is very much more gradual and slower than is observed with estradiol at the menopause in women. Nevertheless, clinically the aging male may, when questioned, admit to occasional episodes of increased heat sometimes associated with sweating. This, however, is not a frequent phenomenon nor so troublesome for the male since there is generally maintenance of reasonable testosterone levels. One should not interpret the relative infrequency of males complaining of such symptoms as evidence that they can surmount the problem better than the female.

It is however of particular interest that we found that men who complain of debilitating episodes of drenching sweats after orchiectomy and who are clearly experiencing hot flushes comparable to those of menopausal women, exhibit a pattern of circulatory change virtually identical to the female flush[25]. Thus there is a rapid initial and marked rise in hand flow which remains elevated for some 2–3 min before falling to baseline (Figure 4). Forearm flow and pulse rate also rise and fall to baseline levels at a time when hand flow is still elevated. In a recent study[26] of men with carcinomatosis of the prostate treated with GnRH agonists, we found overall that there was an increase in mean forearm flow in men after they had been treated with the GnRH agonist and which parallels the rise in forearm flow observed in flushing women.

The possibility that loss of estradiol in the male after gonadectomy or GnRH agonist therapy may be involved in such a response can however be discounted. The occurrence of hot flushes in the female does not, it must be emphasized, correlate with estradiol levels. Flushes, for example, are sometimes most frequent and devastating in the perimenopausal phase when blood estrogen levels fluctuate and can on occasion be 'normal'. Blood estrogen concentration in the male on the contrary is low – only slightly above that of the postmenopausal female. The major change for women at the menopause is loss of circulating estradiol (estrone levels being little reduced) whereas in the male after gonadectomy or GnRH agonist therapy, the major change is that of a reduction in testosterone secretion. Hence, although the vascular response during the hot flush would seem to be identical in the two sexes, the major hormonal change which provides the background for the occurrence of hot flushes is not.

The occurrence of flushes in the male after withdrawal of testosterone, and which are apparently identical to those which occur in women after withdrawal of estradiol, is of considerable interest and relevant to the etiology of the flush in both sexes. That loss of an androgen can activate a flush response identical to that seen after withdrawal of estradiol would imply either that there is a fundamental difference between the hypothalamus of the male and the female – the former responding to testosterone lack and the latter to lack of estradiol – or alternatively

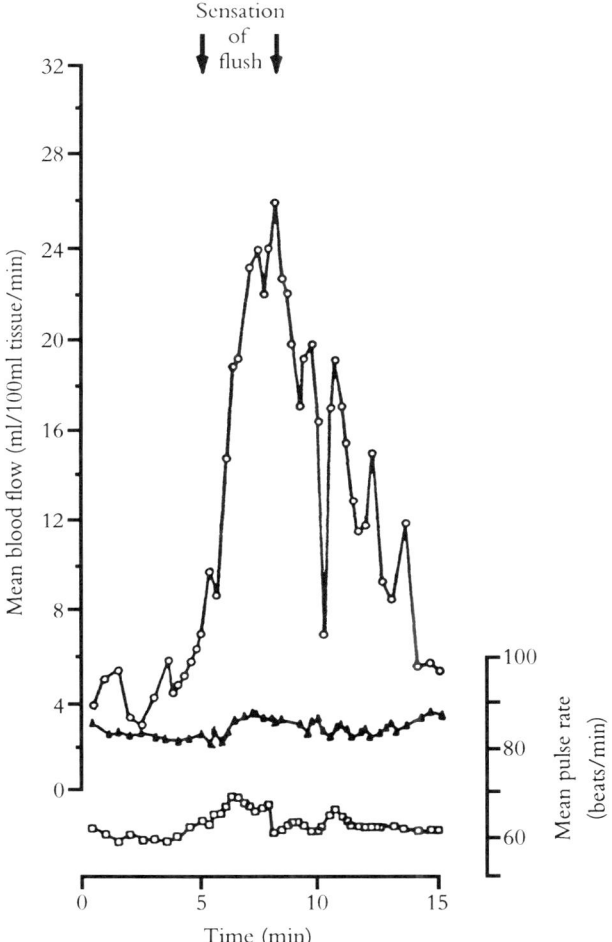

Figure 4 Hand (open circles) and forearm (solid triangles) blood flow and pulse rate (open squares) during a hot flush in a 60-year-old man after orchiectomy. Reproduced with permission from reference 25

that the 'trigger' factor is neither estrogen nor testosterone withdrawal but the loss of some other factor common to both gonads and in whose absence activity of the hypothalamic centers regulating temperature control is disturbed. This in turn may 'fire off' a reflex for releasing and dissipating heat in the absence of a true increase in central core temperature. How this is achieved and the neuronal pathways involved are unknown. The fact that the centers for LH release and temperature regulation are anatomically close has been thought to mean that stimulation of one set of neurons activates the other. This is not necessarily the case in other aspects of hypothalamic function or the activity of other areas of the brain. We need to look for the common factor which links the male and female gonad and circulatory responses. It is clearly not one of the hormones we have studied to date.

Furthermore, the fact that we have now shown that, in the male as well as in the female, those who flush have an elevated forearm flow outside the acute episode of flushing indicates that the flush is only one sign of a more profound disturbance in cardiovascular homeostasis. That this occurs in both sexes and that it has been claimed that flushing women may be more likely to suffer from hypertension and cardiovascular disturbance than asymptomatic women, is another illustration of the fact that the male may also have a problem in this regard. We should pay more serious scientific attention to the nature of the cardiovascular disturbances associated with flushing and not treat them as an irritating social embarrassment for the afflicted individual.

REFERENCES

1. McKinlay, S. M. and Jefferys, M. (1974). The menopausal syndrome. *Br. Med. J.*, **28**, 108–15
2. Lock, M. (1986). Ambiguities of aging: Japanese experience and perceptions of menopause. *Cult. Med. Psychiatry*, **10**, 23–46
3. Beyene, Y. (1986). Climacteric expressions in a crosscultural study. In Nolelovitz, M. and van Keep, P. (eds.) *The Climacteric in Perspective*, pp. 139–47. Proceedings of the Fourth International Congress on the Menopause – Hingham, MA, USA (Lancaster: MTP Press)
4. Molnar, G. W. (1975). Body temperatures during menopausal hot flashes. *J. Appl. Physiol.*, **38**, 499–503
5. Sturdee, D. W., Wilson, K. A., Pipili, E. and Crocker, A. D. (1978). Physiological aspects of menopausal hot flash. *Br. Med. J.*, **2**, 79–80

6. Tataryn, I. V., Meldrum, D. R., Lu, K. H., Frumar, A. M. and Judd, H. L. (1979). LH, FSH and skin temperature during menopausal hot flushes. *J. Clin. Endocrinol. Metab.*, **49**, 152–4

7. Molnar, G. W. (1979). Investigation of hot flashes by ambulatory monitoring. *Am. J. Physiol.*, **6**, R306–10

8. Tataryn, I. V., Lomax, P., Bajorek, J. G., Chesarck, W., Meldrum, D. R. and Judd, H. I. (1980). Postmenopausal hot flushes: a disorder of thermoregulation. *Maturitas*, **2**, 101–7

9. Ginsburg, J., Swinhoe, J. and O'Reilly, B. (1981). Cardiovascular responses during the menopausal hot flush. *Br. J. Obstet. Gynaecol.*, **88**, 925–30

10. Ginsburg, J., Hardiman, P. and O'Reilly, B. (1989). Peripheral blood flow in menopausal women who have hot flushes and in those who do not. *Br. Med. J*, **208**, 1488–90

11. Ginsburg, J. and Hardiman, P. (1990). Cardiovascular effects of transdermal estradiol in postmenopausal women. *Ann. N.Y. Acad. Sci.*, **S92**, 424–5

12. Erlik, Y., Meldrum, D. R. and Judd, H. I. (1982). Estrogen levels in post-menopausal women with hot flashes. *Obstet. Gynecol.*, **59**, 403–7

13. Hagen, C., Christiansen, C., Christensen, M. S. and Transbol, I. (1982). Climacteric symptoms, fat mass and plasma concentrations of LH, FSH, Prl, oestradiol-17B and androstenedione in the early post-menopausal period. *Acta Endocrinol. (Copenh).*, **101**, 87–92

14. Mango, D., Scirpa, P., Battaglia, F. and Bini, E. (1984). Plasma androstenedione and oestrone levels in the climacteric syndrome. *Maturitas*, **5**, 245–50

15. Meldrum, D. R., Tataryn, I. V., Frumar, A. M., Erlik, Y., Lu, K. H. and Judd, H. I. (1980). Gonadotrophins, estrogens and adrenal steroids during the menopausal hot flash. *J. Clin. Endocrinol. Metab.*, **50**, 685–9

16. Genazzani, A. R., Petraglia, P., Facchinetti, I., Facchini, V., Volpe, A. and Alessandrini, G. (1984). Increase of proopiomelanocortin-related peptides during subjective menopausal flushes. *Am. J. Obstet. Gynecol.*, **149**, 775–9

17. Meldrum, D. R., DeFazio, J. D., Erlik, Y., Lu, J. K. H., Wolfsen, A. E., Carlson, H. E., Hershman, J. M. and Judd, H. I. (1984). Pituitary hormones during the menopausal hot flash. *Obstet. Gynecol.*, **64**, 752–6

18. Casper, R. J., Yen, S. S. C. and Wilkes, M. M. (1979). Menopausal flushes; a neuroendocrine link with pulsatile luteinizing hormone secretion. *Science*, **205**, 823–5

19. Gambone, J., Meldrum, D. R., Laufer, I., Chang, R. J., Lu, J. K. H. and Judd, H. I. (1984). Further delineation of hypothalamic dysfunction responsible for menopausal hot flashes. *J. Clin. Endocrinol. Metab.*, **59**, 1097–102

20. Mashchak, C. A., Kletzky, O. A., Artal, R. and Mishell, D. R. Jr. (1984). The relation of physiological changes to subjective symptoms in postmenopausal women with and without hot flushes. *Maturitas*, **6**, 301–8

21. Kronenberg, F., Cole, I. J., Dinkie, D. M., Dyrenfurth, I. and Downey, J. A. (1984). Menopausal hot flashes: thermoregulatory, cardiovascular and circulating catecholamine and LH changes. *Maturitas*, **6**, 31–43

22. Ginsburg, J. and O'Reilly, B. (1987). Are blushes the same as flushes? *Clin. Sci.*, **72** (Suppl. 16), 65

23. Steinfeld, A. D. and Reinhardt, C. (1980). Male climacteric after orchidectomy in patient with prostatic cancer. *Urology*, **16**, 620–2

24. Linde, R., Doette, G. C., Alexander, N., Kirchner, F., Vale, W., River, J. and Rabin, D. (1981). Reversible inhibition of testicular steroidogenesis and spermatogenesis by a potent gonadotropin-releasing hormone agonist in normal men. *N. Engl. J. Med.*, **305**, 663–7

25. Ginsburg, J. and O'Reilly, B. (1983). Climacteric flushing in a man. *Br. Med. J.*, **287**, 262

26. Hardiman, P., Abel, P. D. and Ginsburg, J. (1995). Peripheral vascular effects of gonadotrophin-releasing hormone agonists in men. *Menopause*, **2**, 159–61

DISCUSSION

Gooren: The observations in transsexuals would be consistent with your ideas about a common trigger factor for the occurrence of hot flushes in men and women that is the same for both sexes. In fact, we have seen in transsexuals that it is perfectly possible to suppress hot flushes in females with androgens and in males with estrogens.

Ginsburg: Indeed, you would not regard the hypothalamus of the male and female as different. It seems, however, in view of our respective observations, as if there is some other substance acting between androgens and estrogens.

Gooren: I would follow your assumption that the LH pulse generator has nothing to do with it. If you administer 50 µg of ethinylestradiol to a male-to-female transsexual you do not see a reduction in the pulses, but you see a reduction in the hot flushes. I would also like to ask you the following: do you see a similarity between the appearance of a kind of flush in the upper part of the body of some people after alcohol consumption and the hot flush of the menopause?

Ginsburg: We have not studied alcohol-induced flushing in diabetic subjects taking an oral hypoglycemic agent (a response investigated by others by skin temperature measurements alone) and I do not think it is necessarily the same phenomenon as the menopausal flush. We have, however, investigated facial flushing after nicotinamide administration. We observed a prolonged rise in hand blood flow during the nicotinamide induced flush, a response quite different from our findings during the menopausal flush. I would also like to amplify our observation that the menopausal woman who flushes is very different from the 'blusher', in that the fomer apparently has a circulatory disturbance outside flushing episodes. This would suggest a major change in cardiovascular homeostasis, which may be relevant to subsequent cardiovascular function e.g. whether or not hypertension develops. We are currently following up a cohort of flushing women to see if their future cardiovascular progress differs from non-flushers because of differences in cardiovascular homeostasis. Such changes in vascular setting might be important in terms of control of the circulation and subsequent development – may or may be not – of hypertension and cardiovascular disease.

Bain: Young boys who do not go through puberty because of idiopathic hypogonadotropic hypogonadism do not flush. And one of the reasons might be that until they get some steroid on board, they may not be primed to flushing, whatever the mechanism is. However, many of these young men get treated with testosterone, and if they subsequently stop the treatment because they do not want to have it any more, they still do not flush even when they are off testosterone for some period of time. Now could that be because they have not yet reached an age where the flusing threshold plays a role?

Ginsburg: Frankly, I do not know. It is of course well known that girls with gonadal dysgenesis once they have been treated with estrogens will experience hot flushes if treatment is stopped. I have not however seen enough men who stopped treatment as you described. There are no studies of the vascular responses during flushing after hypophysectomy, so we do not know if the response is the same as that of menopausal flushing. We have, so far, not identified the substance responsible for flusing. I fear we have been measuring too many steroids and known regulators of hormonal pathways, whereas the 'substance' responsible may actually be something quite simple, linked with gonadal failure in both sexes.

9

Androgens, cognitive functioning and mood in men

K. Christiansen

SEX HORMONES AND COGNITION

For many decades, consistent sex differences in tests of cognitive abilities have been widely reported. During certain developmental stages – in particular during the first years of life and again from the 11th year into early adulthood – girls surpass boys in several verbal skills (articulation, verbal understanding, word fluency and perceptual speed). In contrast, males excel after about the 10th year and in adulthood at non-verbal skills, especially at visual–spatial tasks involving spatial rotation and manipulation, at the related concept of field independence, and at mathematical reasoning[1-6]. Numerous sources of variance contribute to the gender differences which have been observed. The relevant factors which have been considered up till now include culture, physical environment, socialization practices, X chromosome linkage of spatial, and partly of verbal, skills as well as the degree of hemispheric lateralization of verbal and non-verbal processes[7-11]. It is also generally accepted that sex hormones play a critical role in sex-typical cognitive abilities as well as in interindividual differences within the sexes. Two decisive phases have been named as the time of the effects of sex hormones on brain structures and consequently upon behavior. During fetal and neonatal life, relatively high concentrations of hormones, especially testosterone, are said to influence brain development by helping to organize the undifferentiated brain in a sex-specific manner. In addition, current sex hormone levels in adult men and women are thought to activate neural structures and hence cognitive abilities.

Clinical studies

Evidence of a connection between sex hormones and spatial abilities came first from studies of phenotypic women with Turner's syndrome (X0 karyotype, no gonadal hormones) or testicular-feminization syndrome (XY karyotype, here the body tissues are unable to respond to normal levels of testosterone). These patients have female external genitalia, they are raised as girls, and develop a feminine gender identity. With regard to cognitive functioning such individuals' verbal skills surpass their spatial abilities, the typical pattern of cognitive functioning in women (Table 1). Moreover, they are also on the average below 'normal' healthy women in terms of their non-verbal performance, but not of their verbal abilities[12–16]. The strikingly poor non-verbal abilities are mainly explained by their low sex hormone levels.

Feminized prepubertal boys suffering from a kwashiorkor-induced endocrine dysfunction have been found to exhibit a more feminine cognitive style than did a male control group. Dawson[17] attributed this finding to a two-way relationship between sex hormones and socialization – because of their physical effeminateness, the boys tended to be raised into more feminine roles.

Studies on men with idiopathic or acquired hypogonadotropic hypogonadism appear to confirm the importance of testosterone for spatial abilities[18,19]. A group of men with idiopathic hypogonadotropic

Table 1   Clinical studies

Sample	Study
Turner's syndrome	Silbert *et al.* (1977)[13]
	Rovet and Netley (1982)[14]
	Dellantonio *et al.* (1984)[15]
Testicular-feminization syndrome	Imperato-McGinley *et al.* (1991)[16]
Feminized prepubertal boys	Masica *et al.*[12]
Hypogonadotropic hypogonadism	Dawson (1972)[17]
	Buchsbaum and Henkin (1980)[18]
Congenital adrenal hyperplasia	Hier and Crowley (1982)[19]
(CAH)	Lewis *et al.* (1968)[20]
	Baker and Ehrhardt (1974)[21]
	McGuire *et al.* (1975)[22]
	Perlman (1973)[23]
	Resnick *et al.* (1986)[24]

hypogonadism, and presumably lifelong testosterone deficiency, were significantly worse than a group of men with late onset of pathologically reduced testosterone or normal controls on a number of spatial tests, but not on verbal tests. As short-term androgen therapy did not restore spatial function, these findings suggest that pre- and perinatal hormonal environments have lifelong effects on intellectual function in humans.

Although controversial, data from children with congenital adrenal hyperplasia (CAH), more or less support this hypothesis. While some surveys[20–22] could not ascertain any effect of the prenatal exposure to higher-than-normal levels of androgens in CAH-children, others[23,24] found that CAH-girls were superior to their siblings or normal controls on spatial tasks. However, in boys with CAH the prenatal increase of androgens did not significantly influence their visual–spatial abilities.

Androgen treatment

Direct manipulation of steroid hormones supports the conclusion that androgens play an important role in cognition. Studies of the effect of hormone treatment on cognitive functioning in adult males date back to 1941 and 1944 when Simonson and his co-workers published their experiments using methyl testosterone (Table 2)[25,26]. Administration of testosterone to castrate human males, eunuchoids and older men improved their ability to perceive flicker (critical flicker frequency), a measure of attention and alertness, as long as the androgen treatment lasted. Düker[27] using either testosterone, estradiol, or a combination of testosterone and estradiol produced a significant rise in concentration and speed in solving simple arithmetical problems in a group of men with severe mental exhaustion. Similar results were reported by Stenn and colleagues[28] who treated three poorly androgenized male adolescents. Intramuscular injections of testosterone enhanced their concentration and performance on a verbal fluency task.

Positive effects of testosterone treatment on cognitive abilities were also observed by Janowsky and co-workers[29]. When normal, aging men were given testosterone to enhance sexual functioning, as a side-effect, they showed improved performance on visual–spatial tests. A similar result was demonstrated when female-to-male transsexuals received high doses of testosterone esters intramuscularly. Their visual–spatial skills improved dramatically and their verbal fluency skills declined considerably within 3

Table 2 Androgen treatment and cognitive abilities in men

Study	Sample	Cognitive task
Simonson et al. (1941)[25]	castrates, eunuchoids	critical flicker frequency⬈
Simonson et al. (1944)[26]	older men	critical flicker frequency⬈
Düker (1957)[27]	men in a state of mental exhaustion	arithmetical problem solving⬈
Klaiber et al (1971)[31]	college students	verbal and non-verbal repetitive tasks⬈
Stenn et al. (1972)[28]	poorly androgenized male adolescents	verbal repetitive tasks⬈
van Goozen et al. (1994)[30]	female-to-male transsexuals	visual–spatial skills⬈ verbal fluency⬊
Janowsky et al. (1994)[29]	older men	visual–spatial ability⬈

⬈increase; ⬊decline

months[30]. Due to ethical reasons, today manipulation of gonadal hormones is restricted to patients in clinical studies. Thus, it is over 20 years ago that Klaiber and colleagues[31] tested effects of infused testosterone on mental performance in healthy male college students. After a 4-h infusion of testosterone or saline infusion in the control group, the postinfusion performances of the control group on a repetitive mental task had a significantly greater decline than in the testosterone-infused group. The result indicates that testosterone acted to prevent the decline from morning to afternoon in performance of simple repetitive verbal or number tasks (the latter decline is a well-known phenomenon in psychology).

Current, endogenous sex hormone levels in normal males

In the early 1970s, determination of sex hormones in biological fluids using radioimmunoassays was well established as a highly specific and reliable method. The possibility of obtaining quantitative data on circulating hormones in blood or saliva initiated a new field of psychobiological research: the relation of sex hormone levels and cognitive abilities in healthy, non-clinical populations, both *intra*individually and *inter*individually. Komnenich and colleagues[32] were first to investigate 10 healthy young men. Four times within a month, they measured the concentration of follicle stimulating hormone (FSH), luteinizing hormone (LH), estradiol, progesterone and testosterone in plasma. On each of the 4 days, simple repetitive verbal tasks and a test of field-independence were administered. The performance on simple verbal repetitive tasks was found to be positively related to estradiol, but none of the other hormones exhibited a significant relation to cognitive performance (Table 3).

In a relatively large sample of men ($n = 43$) between 20 and 40 years a distribution of visual–spatial test scores as a function of androgen levels with the best-fitting third-order polynomial function describing the curve was detected[33]. The authors reported that normal males selected for low plasma androgens were superior on certain spatial tests; while in females, the reverse was true, that is, highest-androgen females were superior to low-androgen women. However, due to the high cross-reactivity of the antibody used in their radioimmunoassays the authors speak of 'general androgen' level instead of free testosterone, which they originally intended to measure. Gouchie and Kimura[34] found an effect similar to that of Shute and co-workers, using not the extremes of the group (of normal men and women) only, but a simple median split to divide all subjects on the basis of saliva testosterone levels. For one of the two spatial tests there was a significant sex-by-hormone level interaction, indicating that low levels of testosterone in males and high levels of testosterone in females are associated with superior performance. On the other hand, multiple regression analyses of the male data alone revealed a trend in men for a positive linear relationship between saliva testosterone levels and a visual–spatial test score ($r = 0.29$, $p = 0.06$).

There was no evidence for a consistent relationship between testosterone concentrations in men or women and performance on perceptual speed tasks or a vocabulary test at which females usually excel.

In contrast to these latter studies[33,34], several others have shown a significant linear hormonal–cognitive relationship. Gordon and Lee[35] investigated 32 young men with a number of visual–spatial and verbal tests and determined serum FSH, LH and testosterone levels of their subjects. Testosterone concentrations correlated significantly positively with a spatial orientation task, but not with any of the spatial or verbal tests.

Christiansen and Knussmann[36] attempted a broader investigation of the effects of sex hormone levels on cognitive abilities, both spatial and non-spatial, in a larger sample of men. From 117 healthy men between the ages of 20 and 30 years blood and saliva samples were collected to determine the serum concentrations of testosterone, 5α-dihydrotestosterone (DHT), and free, non SHBG-bound saliva testosterone. The cognitive functioning

Table 3 Serum and saliva androgens and cognitive abilities in normal males

Study	Cognitive task	Hormonal–cognitive relation
Komnenich et al. (1978)[32]	Verbal	non-significant
	visual–spatial	non-significant
Shute et al. (1983)[33]	visual–spatial	curvilinear
Gordon and Lee (1986)[35]	visual–spatial	positive
Christiansen and Knussmann (1987)[36]	visual–spatial	positive
	field independence	positive
	verbal fluency	negative
Gouchie and Kimura (1991)[34]	visual–spatial	curvilinear
	verbal	non-significant
	perceptual speed	non-significant
Tan and Akgün (1992)[38]	non-verbal	positive
Christiansen (1993)[39]	tactual–spatial	positive
	field independence	positive
	verbal fluency	negative
McKeever and Deyo (1990)[37]	visual–spatial	positive

was ascertained by five spatial and six verbal ipsative test scores, reflecting intraindividual variance in the performance of these tasks, independent of the person's general level of achievement. The relationship between circulating sex hormone levels and cognitive performance exhibited a clear pattern. All of the significant correlations between androgen values and verbal tests were negative. Significant correlations between androgens and spatial and field independence tests were, in contrast, all positive. Correspondingly, a more 'masculine' cognitive pattern (superior skills on spatial tests as compared to verbal tests) was positively related to all three androgen concentrations. It should be noted that total serum testosterone clearly showed the greatest number of significant relations with verbal and spatial test scores. Three years later these findings of a positive correlation between endogenous androgen levels and spatial tasks with regard to dihydrotestosterone were confirmed[37]. Further evidence for a linear androgen–cognitive relationship comes from a Turkish sample of male university students[38] where a positive correlation between non-verbal tasks and testosterone was found. Interestingly, this was only the case for a subsample of right-handed men with right eye preference.

As all the data on the relationship between circulating levels of androgens and cognitive abilities were collected in subjects living in western cultures, Christiansen[39,40] tried to test the validity of previous findings in a non-western group of healthy young males. Altogether 256 !Kung San hunter-gatherers ('bushmen') and Kavango farmers from Namibia/Southern Africa who live mainly at subsistence level in their traditional lifestyle with a low degree of transition to western culture were investigated. Blood and saliva samples were collected in order to determine total testosterone, DHT and estradiol in the serum and 'free', salivary testosterone. In order to make possible a comparison of previous findings of hormone-related variations in cognitive performance, tactual–spatial, field-independence, and verbal tests were the same or similar to those used previously[36], adapted however for the testing of illiterate subjects with no experience in paper-and-pencil tasks. In spite of the problems inherent in cross-cultural psychological research such as item bias, construct validity, functional, conceptual and metric equivalence of tests, the African data yielded the same hormonal-cognitive pattern as was found in western samples. Total and salivary testosterone showed the greatest number of significant relationships, a positive one with visual or tactual spatial tasks and a negative correlation with verbal tests.

The short list of biological underpinnings of sex hormone effects on cognitive abilities makes it reasonable to conclude that androgens play an activating role in cognitive functioning throughout life – from prenatal life through adulthood till old age. But it has to be noted that explanations of inter- and intraindividual differences in cognitive abilities are complex and any causal model will have to recognize the reciprocal effects that environment and biology have on each other.

MOOD

Affective disorders, in particular depression, have always been the focus of psychiatric investigations because depression is a prominent illness. It is widely acknowledged that depressive illness comprises a heterogeneous group of disorders that have different clinical symptom pictures and that respond to different treatment modalities. The symptoms of depression include reduced mood (profound sadness); low self-esteem and feelings of worthlessness; general fatigue; feelings of guilt; disturbances, usually reductions, in sleep, sex drive and food intake; anger; absence of pleasure; and agitated or retarded motor symptoms (DSM-III Diagnostic and Statistical Manual of Mental Disorders, American Psychiatric Association). Thus, depression covers a wide range of emotional and clinical states, and as a normal mood, depression is ubiquitous in human existence. While classification of affective disorders is controversial[41,42], it is an uncontradicted finding that depression is more prevalent in women than in men. This phenomenon could be explained in part by the hypothesis that sex steroid hormones are involved in the etiology of some types of depression[43]. The relationship between cyclic hormonal changes associated with the menstrual cycle and changes in mood and social behavior has been examined in great detail[44-47]. However, the interrelationships between affective states in men and their corresponding sex hormone level is an area of research in which little work has been reported.

The study of testosterone in depression was prompted by the observation that in castrates and hypogonadal men treated with androgen the incidence of depressed mood and emotional instability appreciably decreased. Beach[48] reported on several cases of involutional melancholia in older men which were explained by a reduction of testicular hormones reduced below the level necessary for mental stability. Improvement in mood and alertness was described in 65% of such cases when treated with testosterone propionate. However, more recent physiologic approaches to the psychoendocrinology of depression, using radioimmunoassays for the determination of androgen levels in body fluids, were less successful in demonstrating anti-depressive properties of endogenous testosterone in male patients (Table 4). Only Vogel[49] and Yesavage[50] and their collaborators found a significantly negative relation of testosterone with depression. In comparison to healthy controls, their patients had 30% lower levels of total and free testosterone, while estradiol showed an increase of about 50%. In the sample of Yesavage, severity of depressive symptoms correlated significantly negatively with testosterone[50]. The majority of clinical studies failed to ascertain significant effects of testosterone, testosterone production rate, LH, FSH and estradiol on depression[51-6]. Although some data suggested a dysfunction of the hypothalamo–pituitary–gonadal (HPG) axis, the majority of researchers concluded that there appears to be no major dysregulation of HPG axis activity in male endogenous depressives, even if dysregulation might be related to the reduced energy levels and sexual appetite occurring in many depressive men[56-60].

Table 4 Endogenous androgens and mood: clinical studies

Study	Effect of testosterone on depression
Persky *et al.* (1968)[51]	NS
Persky *et al.* (1971)[52]	NS
Sachar *et al.* (1973)[53]	NS
Vogel *et al.* (1978)[49]	elucidation
Rubin *et al.* (1981)[54]	NS
Yesavage *et al.* (1985)[50]	elucidation
Rubin *et al.* (1989)[56]	NS

Parallel to clinical studies on male depressives, research was extended to sex hormone levels and mood in healthy males without symptoms of severe depression. Surprisingly, here quite a number of significant, however controversial, relationships between mood and sex hormone status could be demonstrated, although hormonal and psychological variables lay within the normal range. In a 2-month study of testosterone cycles and affective states among 20 healthy young males, Doering and collaborators[61] observed significant positive correlations between testosterone and self-ratings of depression (Table 5). Houser[62] tested five healthy young men three times a week over a 10-week period and observed a significantly positive relationship between testosterone and self-ratings of depression with the Multiple Affect Adjective Check List and correspondingly a negative correlation with the elation scale. In a group of adolescent boys between 9 and 14 years of age positive relationships of DHEA, an adrenal androgen, and estradiol with self-ratings of sadness were found[63].

Christiansen and Knussmann (unpublished data) confirmed these findings with a corresponding trend in a large sample ($n = 117$) of unselected males. Men with a high testosterone level had higher depression scores ($p < 0.06$) on a self-rating scale than low testosterone males.

On the other hand, there is also some evidence for an association of high testosterone levels with emotional well-being as it has been found in early pharmacological research. A sample of 21 healthy young males was investigated and report was made of significantly negative correlations of saliva testosterone with depression and anxiety[64] (compare refs. 65 and 66) and a positive correlation with joyfulness. Research on anabolic steroid abuse supported these findings. Researchers contacted body building studios and offered members using steroids a cash payment to engage in confidential interviews about their steroid use. Thirteen of 41 individuals reported to be

Table 5 Androgens and mood in the normal male

Study	Relationship of androgens and self-ratings of mood
Doering *et al.* (1974)[61]	positive with depression
Houser (1979)[62]	positive with depression
	negative with elation
Christiansen *et al.* (1984)[65]	negative with anxiety
Susman *et al.* (1987)[63]	positive with depression
Diamond *et al.* (1989)[66]	negative with anxiety
Hubert (1990)[64]	positive with joyfulness
	negative with anxiety
	negative with depression
Christiansen and Knussmann (unpublished results)	positive with depression

manic or near manic. Most symptoms subsided when anabolic steroid use was discontinued[67]. According to Kashkin and Kleber[68] anabolic steroids might have direct reward properties. Increasing reports of unexpected suicides in previously non-depressed young men who abruptly stopped using anabolic steroids have been noted. Thus, the withdrawal symptoms manifested by anabolic steroid abusers and the symptoms of premenstrual syndrome-related or postpartum depression may result from a common underlying cause, namely, dependence upon elevated steroid hormone levels[42].

There are several explanations for these contradictory results of significantly positive and negative correlations of testosterone and depression. Differences in selection of samples, personality tests, sampling procedures and determinations of sex hormones, and sometimes very small samples have to be considered. Houser wrote in 1979: 'The findings . . . provide sufficient encouragement for future work, and the evidence is strong enough to warrant further investigation of . . . an affect–endocrine relationship'[62]. Even 16 years later, in 1995, there is still a lot to do.

REFERENCES

1. Maccoby, E. E. and Jacklin, C. N. (1974). *The Psychology of Sex Differences.* (Stanford: Stanford University Press)

2. Wittig, M. C. and Petersen, A. C. (1979). *Determinants of Sex-Related Differences in Cognitive Functioning*. (New York: Academic Press)

3. Newcombe, N., Bandura, M. M. and Taylor, D. G. (1983). Sex differences in spatial ability and spatial activities. *Sex Roles*, **9**, 377–85

4. Hyde, J. S. and Linn, M. C. (1988). Gender differences in verbal ability: a meta-analysis. *Psychol. Bull.*, **104**, 53–69

5. Hyde, J. S., Fennema, E. and Lamon, S. J. (1990). Gender differences in mathematics performance: a meta-analysis. *Psychol. Bull.*, **107**, 139–55

6. Halpern, D. F. (1992). *Sex Differences in Cognitive Abilities*. (Hillsdale, New Jersey: Lawrence Erlbaum Associates)

7. Waber, D. P. (1977). Sex differences in mental abilities, hemispheric lateralization, and rate of physical growth at adolescence. *Develop. Psychol.*, **13**, 29–38

8. Vandenberg, S. G. and Kuse, A. R. (1979). Spatial ability: a critical review of the sex-linked major gene hypothesis. In Wittig, M. C. and Petersen, A. C. (eds.) *Determinants of Sex-related Differences in Cognitive Functioning*, pp. 67–95. (New York: Academic Press)

9. Blatter, P. (1982). Sex differences in spatial ability: the X-linked gene theory. *Percep. Mot. Skills*, **55**, 455–62

10. Hahn, W. K. (1987). Cerebral lateralization of function. From infancy through childhood. *Psychol. Bull.*, **101**, 376–92

11. Ashton, G. C. and Borecki, I. B. (1987). Further evidence for a gene influencing spatial ability. *Behav. Genet.*, **17**, 243–56

12. Masica, D. N., Money, J., Ehrhardt, A. A. and Lewis, V. G. (1969). IQ, fetal sex hormones and cognitive patterns: studies in the testicular feminizing syndrome of androgen insensitivity. *Johns Hopkins Med. J.*, **124**, 33–43

13. Silbert, A. R., Wolff, P. H. and Lilienthal, J. (1977). Spatial and temporal processing in patients with Turner's syndrome. *Behav. Genet.*, **7**, 11–21

14. Rovet, J. and Netley, C. (1982). Processing deficits in Turner's syndrome. *Develop. Psychol.*, **18**, 77–94

15. Dellantonio, A., Lis, A., Saviolo, N., Rigon, F. and Tenconi, R. (1984). Spatial performance and hemispheric specialization in the Turner syndrome. *Acta Med. Auxol.*, **16**, 193–203

16. Imperato-McGinley, J., Pichardo, M., Gautier, T., Voyer, D. and Bryden, M. P. (1991). Cognitive abilities in androgen-insensitive subjects: comparison with control males and females from the same kindred. *Clin. Endocrinol.*, **34**, 341–7

17. Dawson, J. L. M. (1972). Effects of sex hormones on cognitive style in rats and men. *Behav. Genet.*, **2**, 21–42
18. Buchsbaum, M. S. and Henkin, R. I. (1980). Perceptual abnormalities in patients with chromatin negative gonadal dysgenesis and hypogonadotropic hypogonadism. *Int. J. Neurosci.*, **11**, 201–9
19. Hier, D. B. and Crowley, W. F. (1982). Spatial ability in androgen-deficient men. *N. Engl. J. Med.*, **306**, 1202–5
20. Lewis, V. G., Money, J. and Epstein, R. (1968). Concordance of verbal and nonverbal ability in the adrenogenital syndrome. *Johns Hopkins Med. J.*, **122**, 192–5
21. Baker, S. W. and Ehrhardt, A. A. (1974). Prenatal androgen, intelligence, and cognitive sex differences. In Friedman, R. C., Richart, R. M. and Vande Wiele, R. L. (eds.) *Sex Differences in Behavior*, pp. 53–76. (New York: Wiley)
22. McGuire, L. S., Ryan, K. O. and Omenn, G. S. (1975). Congenital adrenal hyperplasia: II. Cognitive and behavioral studies. *Behav. Gen.*, **5**, 175–88
23. Perlman, S. M. (1973). Cognitive abilities of children with hormone abnormalities: screening by psychoeducational tests. *J. Learn. Disabil.*, **6**, 21–9
24. Resnick, S. M., Berenbaum, S. A., Gottesman, I. I. and Bouchard, T. J. Jr. (1986). Early hormonal influences on cognitive functioning in congenital adrenal hyperplasia. *Develop. Psychol.*, **22**, 191–8
25. Simonson, E., Kearns, W. M. and Enzer, N. (1941). Effect of oral administration of methyltestosterone on fatigue in eunuchoids and castrates. *Endocrinology*, **28**, 506–12
26. Simonson, E., Kearns, W. M. and Enzer, N. (1944). Effect of methyl testosterone treatment on muscular performance and the central nervous system of older men. *J. Clin. Endocrinol.*, **4**, 528–34
27. Düker, H. (1957). *Leistungsfähigkeit und Keimdrüsenhormone.*(München: Barth)
28. Stenn, P. O., Klaiber, E. L., Vogel, W. and Broverman, D. M. (1972). Testosterone effects on photic stimulation of the EEG and mental performances of humans. *Percept. Mot. Skills*, **34**, 371–8
29. Janowsky, J. S., Oviatt, S. K. and Orwoll, E. S. (1994). Testosterone influences spatial cognition in older men. *Behav. Neurosci.*, **108**, 325–32
30. van Goozen, S. H. M., Cohen-Kettenis, P. T., Gooren, L. J. G., Frijda, N. H. and van de Poll, N. E. (1994). Activating effects of androgens on cognitive performance: causal evidence in a group of female-to-male transsexuals. *Neuropsychologia*, **32**, 1153–7

31. Klaiber, E. L., Broverman, D. M., Vogel, W., Abraham, G. E. and Cone, F. L. (1971). Effects of infused testosterone on mental performances and serum LH. *J. Clin. Endocrinol. Metab.*, **32**, 341–9

32. Komnenich, P., Lane, D. M., Dickey, R. P. and Stone, S. C. (1978). Gonadal hormones and cognitive performance. *Physiol. Psychol.*, **6**, 115–20

33. Shute, V. J., Pellegrino, J. W., Hubert, L. and Reynolds, R. W. (1983). The relationship between androgen levels and human spatial abilities. *Bull Psychonomic Soc.*, **21**, 465–8

34. Gouchie, C. T. and Kimura, D. (1991). The relation between testosterone levels and cognitive ability patterns. *Psychoneuroendocrinology*, **16**, 323–34

35. Gordon, H. W. and Lee, P. A. (1986). A relationship between gonadotropins and visuospatial function. *Neuropsychologia*, **24**, 563–76

36. Christiansen, K. and Knussmann, R. (1987). Sex hormones and cognitive functioning in men. *Neuropsychobiology*, **18**, 27–36

37. McKeever, W. F. and Deyo, R. A. (1990). Testosterone, dihydrotestosterone, and spatial task performances of males. *Bull. Psychonomic Soc.*, **28**, 305–8

38. Tan, Ü. and Akgün, A. (1992). There is a direct relationship between nonverbal intelligence and serum testosterone level in young men. *Int. J. Neurosci.*, **64**, 213–6

39. Christiansen, K. (1993). Sex-hormone related variations of cognitive performance in !Kung San hunter-gatherers of Namibia. *Neuropsychobiology*, **27**, 97–107

40. Christiansen, K. (1996). The role of sex hormones, diet and acculturation for the psychological differentiation of hunter-gatherers (San) and sedentary farmers (Kavango) of Namibia/Southern Africa. *J. Cross-cult. Psychol.* (in press)

41. Klerman, G. L. (1983). The scope of depression. In Angst, J. (ed.) *The Origins of Depression: Current Concepts and Approaches*, pp. 5–25. (Berlin: Springer)

42. Nelson, R. J. (1995). *An Introduction to Behavioral Endocrinology*. (Sunderland, MA: Sinauer Assoc.)

43. Maggi, A. and Perez, J. (1985). Role of female gonadal hormones in the CNS: clinical and experimental aspects. *Life Sci.*, **37**, 893–906

44. Bains, G. K. and Slade, P. (1988). Attributional patterns, moods, and the menstrual cycle. *Psychosom. Med.*, **50**, 469–76

45. Bäckström, T., Sanders, D., Leask, R., Davidson, D., Warner, P. and Bancroft, J. (1983). Mood, sexuality, hormones, and the menstrual cycle. II. Hormone

levels and their relationship to the premenstrual syndrome. *Psychosom. Med.*, **45**, 503–7

46. Lewis, J. W. (1990). Premenstrual syndrome as a criminal defense. *Arch. Sex. Behav.*, **19**, 425–41

47. Sherwin, B. B. (1991). The impact of different doses of estrogen and progestin on mood and sexual behavior in postmenopausal women. *J. Clin. Endocrinol. Metab.*, **72**, 336–43

48. Beach, F. A. (1948). *Hormones and Behavior.* (New York: Hoeber)

49. Vogel, W., Klaiber, E. L. and Broverman, D. (1978). Roles of the gonadal steroid hormones in psychiatric depression in men and women. *Prog. Neuro-Psychopharmacol.*, **2**, 487–503

50. Yesavage, J. A., Davidson, J., Widrow, L. and Berger, P. A. (1985). Plasma testosterone levels, depression, sexuality, and age. *Biol. Psychiatry*, **20**, 222–5

51. Persky, H., Zuckerman, M. and Curtis, G. C. (1968). Endocrine function in emotionally disturbed and normal men. *J. Nervous Mental Dis.*, **146**, 488–97

52. Persky, H., Smith, K. D. and Basu, G. K. (1971). Relation of psychological measures of aggression and hostility to testosterone production in man. *Psychosom. Med.*, **33**, 265–77

53. Sachar, E. J., Halpern, F., Rosenfeld, R. S., Gallagher, T. F. and Hellman, L. (1973). Plasma and urinary testosterone levels in depressed men. *Arch. Gen. Psychiatry*, **28**, 15–18

54. Rubin, R. T., Poland, R. E., Tower, B. B., Hart, P. A., Blodgett, A. L. N. and Forster, B. (1981). Hypothalamo–pituitary–gonadal function in primary endogenously depressed men: preliminary findings. In Fuxe, K., Gustafsson, J. A., Wetterberg, L. (eds.) *Steroid Hormone Regulation of the Brain*, pp. 387–96. (Oxford: Pergamon)

55. Unden, F., Ljunggren, J. G., Beck-Friis, J., Kjellman, B. F. and Wetterberg, L. (1988). Hypothalamic–pituitary–gonadal axis in major depressive disorders. *Acta Psychiatr. Scand.*, **78**, 138–46

56. Rubin, R. T., Poland, R. E. and Lesser, I. M. (1989). Neuroendocrine aspects of primary endogenous depression VIII. Pituitary–gonadal axis activity in male patients and matched control subjects. *Psychoneuroendocrinology*, **14**, 217–29

57. Matussek, N. (1980). Zur Biochemie und Neuroendokrinologie der Depression. In Heimann, H. and Giedke, H. (eds.) *Neue Perspektiven in der Depressionsforschung*, pp. 64–72. (Bern: Huber)

58. Freedman, R. and Carter, D. B. (1982). Neuroendocrine strategies in psychiatric research. In Vernadakis, A. and Timiras, P. S. (eds.) *Hormones in Development and Aging*, pp. 619–36. (Lancaster: MTP Press)

59. Angst, J. (1983). *The Origins of Depression: Current Concepts and Approaches.* (Berlin: Springer)

60. Von Zerssen, D., Berger, M. and Doerr, P. (1984). Neuroendocrine dysfunction in subtypes of depression. In Shah, N. S. and Donald, A. G. (eds.) *Psychoneuroendocrine Dysfunction in Psychiatric and Neurological Illnesses: Influence of Psychopharmacological Agents*, pp. 357–82. (New York: Plenum)

61. Doering, C. H., Brodie, H. K. H., Kraemer, H., Becker, H. and Hamburg, D. A. (1974). Plasma testosterone levels and psychologic measures in men over a 2-month period. In Friedman, R. C., Richart, R. M., Van de Wiele, R. L. and Stern, L. O. (eds.) *Sex Differences in Behavior*, pp. 413–32. (New York: Wiley)

62. Houser, B. B. (1979). An investigation of the correlation between hormonal levels in males and mood, behavior and physical discomfort. *Horm. Behav.*, **12**, 185–97

63. Susman, E. J., Inoff-Germain, G., Nottelmann, E. D., Loriaux, D. L., Cutler, G. B. Jr. and Chrousos, G. P. (1987). Hormones, emotional dispositions, and aggressive attributes in young adolescents. *Child Develop.*, **58**, 1114–34

64. Hubert, W. (1990). Psychotropic effects of testosterone. In Nieschlag, E. and Behre, H. M. (eds.) *Testosterone: Action, Deficiency, Substitution.* pp. 51–71. (Berlin: Springer)

65. Christiansen, K., Knussmann, R. and Couwenbergs, C. (1984). Zusammenhänge zwischen Sexualhormonen des Mannes und Ernährung, Streβ und Sexualverhalten. *Homo*, **35**, 251–72

66. Diamond, P., Brisson, G. R., Caudas, B. and Péronnet, F. (1989). Trait anxiety, submaximal exercise and blood androgens. *Eur. J. Applied Physiol. Occup. Physiol.*, **58**, 699–704

67. Pope, H. G. and Katz, D. L. (1989). Homicide and near-homicide by anabolic steroid users. *J. Clin. Psychiatry*, **51**, 28–31

68. Kashkin, K. B. and Kleber, H. D. (1989). Hooked on hormones? An anabolic steroid addiction hypothesis. *J. Am. Med. Assoc.*, **262**, 3166–70

DISCUSSION

Gooren: One of our female to male transsexuals wrote poetry before androgen treatment. After the treatment, she could not find the words for poetry any more. Observations in transsexuals have led us to conclude that you can only be good at one faculty: if you gain in verbal fluency with estrogens, you loose spatial ability, whereas with androgens you may gain spatial ability but loose verbal fluency. We found that androgenised females performed almost the same as normal eugonadal males, and testosterone deprived males performed in the same range as females. Therefore, it looks very much as if your actual levels of sex steroids determine your spatial ability and verbal fluency.

Tenover: After what period of time did you see these changes when you changed hormone levels? Was it very rapid? Were they maintained over a long period of time?

Gooren: We did out measurements after four months of treatment, and we have, so far, not yet done a follow-up.

Orwoll: In animals aromatase activity is diffusely distributed in the brain. Many sex-specific behaviors can be attributed to aromatase activity and some androgen effects are mediated by their conversion to estrogens in certain areas of the brain. Do you know of any data that would indicate that non-aromatisable androgens have the same effects on cognition as aromatis-able androgens?

Gooren: I am not aware of such data in as much as the studies have to my knowledge not been performed.

Morales: In our clinical experience we have treated a number of hy-pogonadic men with injectable esters. They frequently reported a 'roller-coaster' effect. Their sensation of well-being changed by the second week. By use of visual analogue scales to assess mood changes, we noticed that oral androgens more often resulted in a very significant improvement of mood.

Oddens: Do you have any comments on that, Dr Christiansen, because you described so very well that the results of androgens with respect to mood are sometimes negative and sometimes positive?

Christiansen: In hypogonadal men it is always a possibility that androgen administration raises mood. However, in our studies on normal healthy men inter-individual variability plays a greater role. You see quite often – and this I think explains the contrasting results – that due to heterogeneity of individual life-styles and subsequently different exogenous factors influencing testosterone levels among the men studied, significant positive correlations and negative correlations with testosterone emerge.

Tenover: In the supplementation study that we did among elderly men with low androgen levels[1] the subjects were not selected for any kind of behavioral problem or sexual dysfunction. We measured mood and sexual function at baseline and during treatment. When they were on testosterone, mood improved and sexual interest and activity increased (although to a variable extent) in 12 out of the 13 men, as compared with the placebo phase. In a current larger study in which the men are blinded to treatment, I am not seeing the same sort of magnitude of effects that I did in the earlier study. This might be related to the fact that in the current study the men are on treatment for well over two years, and they might not be able to remember what it was like before they began therapy, when they were not being treated.

de Wied: Dr Christiansen, you did not measure learning and memory capacities of the subjects in your studies, did you? I thought you might have possibly adopted a quite narrow focus by looking at spatial and verbal abilities, whilst you know already that men are doing better on spatial and women on verbal tasks. In the perspective that these types of behaviors can apparently be easily affected by administration of hormones, as Dr Gooren described, why did you not choose tests that would examine learning and memory *per se*?

Christiansen: In fact, we did not measure learning and memory, and nobody else does it either. I agree this is a field where much work still needs to be done. I am convinced that there should be effects in this connection, because sex hormones are supposed to raise neuronal activation.

Bain: One thing that we did not speak about so far is the reported aggression in men with testosterone treatment. There is the popular perception that testosterone treatment induces aggression and I have yet to be convinced that this is in fact the case. We have been involved over the years in studies of men who have engaged in sexually aggressive behavior and we have done broad hormonal studies, but we have not found any differences in testosterone levels or other hormones for that matter. Furthermore, in the literature one does not really find good evidence that there is any relationship between serum testosterone and current aggressive behavior.

Gooren: In female to male transsexuals given testosterone we have observed increases in imagined aggression, but not in overt aggression.

Schiavi: John Money[2] has advocated the use of anti-androgens for the treatment of rapists and other paraphiliac behaviors. These behaviors are not considered as sexual antisocial behavior but as aggressive behavior, and significant therapeutic effects have been noted. This would suggest that at least anti-androgens, however their function is mediated, have an effect on such behaviors. Of course these situations are atypical, and we have to be cautious about generalising them.

Bain: I would agree that when you give such men anti-androgens, you can significantly ameliorate their behavior, whatever anti-androgen approach you use. You would think that one side of the coin matches the other side, but I do not think there is evidence to suggest that.

Christiansen: There is quite consistent evidence that a very small portion, however significant, of testosterone, explains aggressive behavior, or better said, the personality trait of aggressiveness. But the proportion is very small; about 5 to 9% of the total variance of the behavior can be explained by testosterone and not more. However, you will not find it in a population of sex offenders, because this is a very special population, with very special problems.

Tenover: I really wonder whether these findings about aggressive personality are at any level clinically significant in terms of androgen replacement therapy.

Gooren: The studies we have done in transsexuals have shown there may be effects, but these are not very large and not really clinically significant. The same can be said about verbal fluency, as I illustrated by the example of this one poet.

REFERENCES

1. Tenover, J. S. (1992). Effects of testosterone supplementation in the aging male. *J. Clin. Endocrinol. Metab.*, **75**, 1092–8

2. Money, J. (1977). Paraphilias. In: Money, J. and Musaph, H. (eds.) *Handbook of Sexology*, pp. 917–28 (Amsterdam: Elsevier/North Holland Biomedical Press)

10

Effects of androgens in women

B. J. Oddens, W. E. Bergink and H. M. Vemer

INTRODUCTION

Androgens (Greek: *andro-* = male, *gennan* = to produce, thus to confer masculinizing and virilizing properties) are produced in both males and females, albeit in women to a lesser extent than in men. The major circulating androgen in males is dihydrotestosterone whereas in females principally testosterone is the active hormone. Testosterone in females is produced by the ovaries (25%), the adrenal glands (25%) and by peripheral conversion (50%) of the prehormones androstenedione and its precursor dehydroepiandrosterone (DHEA). Androstenedione is produced by the ovaries (50%) and adrenals (50%). DHEA and its sulfate (DHEAS) are almost exclusively produced by the adrenals (90–95%)[1]. DHEA and DHEAS are converted in the ovaries[2] and other tissues to testosterone and to prehormones of testosterone with some estrogenic properties like Δ5-androstenediol[3]. Since in females the peripheral conversion of the various prehormones is so important, not only serum testosterone but in particular peripheral conversion of its precursors determine the physiological effects of androgens in women. It must be noted that testosterone and its precursor androstenedione are in turn aromatizable and can be converted into 17β-estradiol and estrone, respectively.

Androgens have been found to convey a wide variety of effects in women. Androgens are involved in sexual differentiation and behavioral imprinting during early fetal development. The latter illustrates the organizational (structural) effects of androgens with respect to the central nervous system, which have also been suggested to play a role in later life. Activational effects have been observed during puberty and in adult women. For example, the variable androgen production during the menstrual cycle seems to influence behavior in women. Effects of hyperandrogenism, such as in the polycystic ovary syndrome, have been clearly documented.

Following androgen supplementation to ovariectomized women, interesting effects have been noted with respect to sexual and psychological functioning, as well as bone metabolism. Possible effects of androgen treatment in postmenopausal women with intact ovaries might warrant attention. These various issues will be discussed in the present chapter.

ORGANIZATIONAL AND ACTIVATIONAL EFFECTS OF ANDROGENS

Phenotypic sex

Classically, it has been known that the absence or presence of androgens will determine the sexual differentiation of the early embryo resulting in the phenotypic sex. Male and female human embryos develop identically until week 8 of gestation. Between week 8 and week 12 sexual differentiation takes place that is marked in the female embryo by degeneration of the mesonephric (Wolffian) ducts and development of the paramesonephric (Müllerian) ducts whereas in male embryos the Wolffian ducts will develop and the Müllerian ducts will degenerate. The paramesonephric ducts will form the female genital tract.

This differentiation depends on the absence or presence of androgens. Feminization of the genital tract results from the absence of androgens or from androgen insensitivity; the fetal ovaries have probably no role in determining sexual differentiation.

Central nervous system: early development

Hormonal influences in early development not only determine the phenotypic sex but may also 'program' the brain to function according to typical patterns of a particular sex.

Consistent differences have been noticed in cognitive functioning between women and men, for example with respect to verbal skills and spatial abilities. These differences might be related to a wide variety of factors[4], which not only include culture, physical environment, socialization practices, X chromosome linkage of these skills and their degree of hemispheric lateralization, but also hormonal factors[5]. Studies in men and women with congenital disorders resulting in low circulating androgen levels or insensitivity to androgens suggested that androgen exposure during early, even fetal, development, may affect gender-typical cognitive behaviors expressed in later life[4]. In this connection, it has been found that gonadal steroids, including androgens, were able to alter the structure of brain cells[5–7]. Data from a variety of non-human species have indicated that early hormones had permanent sexually differential effects on the development of the brain, as well as on subsequent learning abilities[5]. Gender-linked cognitive dimorphism could be directly related to dimorphism in the brain[8]. Some of the sex hormone-related changes were reversible (activational) whereas others were permanent (organizational[5] or structural[9]). In general it appeared that gender differences in this respect were the product of both the imprinting action of gonadal hormones during prenatal and early postnatal development, and activational effects expressed during puberty and in adult life.

Early effects of gonadal hormones on behavior in later life may reach further than cognitive functioning. Udry and co-workers studied a wide variety of gendered behavior in 250 daughters (aged 27–30) of women whose pregnancy serum samples had been stored in 1960[10]. In this way, they were able to study relationships between the scores of the daughters on sex-dimorphic behavior scales, their current androgen levels and the androgen level of their mothers at the time they were pregnant with these daughters (as a proxy for fetal exposure to androgens). It emerged that more feminine patterns of behavior were negatively related to adult androgen levels and positively to maternal sex hormone-binding globulin (SHBG) levels during the second trimester (Table 1; results relating to other trimesters during pregnancy were not significant). Higher SHBG levels represent a higher capacity for binding androgens (resulting in lower levels of bioactive or free androgens). According to the authors the conclusion seemed warranted that the degree of SHBG binding of maternal androgens, and thus the fetal exposure to bioactive androgens, significantly predicted future sex-dimorphic behaviors. Consistent with the above-mentioned notion that early imprinting may interact with activational effects in later life, they found a significant interaction between the effects of prenatal–maternal androgen levels and adult–daughter levels on the behaviors studied. In other words, this suggested that higher fetal exposure to androgens may result in greater responsiveness to adult androgens with respect to gendered behavior in later life.

Table 1 Analysis of gendered behavior on hormonal factors

	Correlations Gendered[†]	Regression Gendered	Interviewer's femininity impression[‡]
Adult testosterone (T1)	−0.20**	−0.22**	—
Adult androstenedione (A)	−0.25**	§	−0.09
Adult SHBG	0.12*	0.04	0.21**
Prenatal second trimester testosterone (T2)	0.01	−0.06	0.11
Prenatal second trimester SHBG	0.29**	0.31**	0.14*
Interaction T1 × T2		0.19*	—
Interaction A × T2		—	0.14*
R^2	—	0.18	0.17

*$p < 0.05$; **$p < 0.01$
[†]Gendered, overall measure of gendered behavior derived from 19 measurement scales relating to various behavioral components (higher score indicating more feminine behavior).
[‡]Scored on a five-point scale ranging from masculine to feminine.
§Model including adult androstenedione instead of adult testosterone was nearly identical.
Data from ref. 10

Central nervous system: advanced effects

As mentioned, gonadal hormones also appeared to affect central nervous system functioning in postnatal life. Both estrogens and androgens have been associated with organizational and activational effects on the brain.

Estrogen administration in rats resulted in enhancement of spatial memory performance[11]. These changes in performance are consistent with structural changes of specific cells observed within the hippocampus[12]. Women usually perform better than men in verbal ability, perceptual speed and accuracy and fine motor skills[5,13] and evidence supports the contention that estrogens influence these aspects of cognitive functioning[14] not only during the fertile period of their life, but also as a result of estrogen replacement therapy[15–17]. In this connection, it has been speculated that estrogens directly affect choline acetyltransferase expression and the activity of acetylcholine receptors in basal forebrain cholinergic neurons[18,19]. Furthermore, they may modulate neuron loss in the medial preoptic nucleus[20] and regulate synaptogenesis in the CA1 region of the hippocampus[19].

The effects of androgens have been less extensively documented. In men, androgens were reported to be negatively correlated with some features of verbal skills[4]. On the other hand, androgens may improve spatial memory abilities and mathematical reasoning[4,5]. A therapeutic role for androgens in enhancing the reparative response of motor neurons to injuries has been proposed[21]. In subhuman primates[22] and other vertebrates, such as female rats, treated with androgens, effects of androgens on learning abilities have been described. These female rats, for example, performed quicker learning on the radial-arm-maze test than untreated controls[23,24], consistent with organizational differences in the brain[22,24]. However, the effects of confounding by differences in levels of anxiety or exploration cannot be discarded in these studies and it remains to be investigated whether the observations are relevant to human cognitive performance[25,26]. Interestingly, perinatal treatment of rats with androgens resulted in permanent changes in the levels of steroid-metabolizing liver enzymes[27]. Therefore, it has been suggested that perinatal exposure to androgens may modify the levels of bioavailable androgens in later life, and in this way the expression of androgenic behavioral and cognitive effects.

ANDROGENS AND THE MENSTRUAL CYCLE

Androgens are produced under the influence of luteinizing hormone (LH) by the theca and interstitial stroma cells of developing follicles and converted into estrogens under the influence of follicle-stimulating hormone (FSH) by granulosa cell aromatase activity[28]. Also, the growth and further development of follicles during fertile life is under the influence of FSH. When at the end of the follicular phase the amount of FSH decreases, the follicle with most FSH receptors has a significant advantage in continuing on its course of development (dominant follicle) while remaining follicles are arrested in their development[29]. In the remaining follicles, insufficient FSH stimulation results in a decrease of conversion to estrogens and thus an increase of androgen secretion. Judd and Yen showed that androgen levels are highest during the middle third of the menstrual cycle[30].

Higher androgen levels in combination with insufficient FSH stimulation will result in a 'wave of atresia'. This can be quantified as an apoptotic process by gel fractionation or histological estimation of the internucleosomal degradation of DNA[31]. Furthermore, androgens stimulate inhibin secretion by the dominant follicle, which further reduces FSH secretion. Follicular testosterone has been suggested to be involved as a local regulating factor in the selection of the dominant follicle[32]. Non-aromatizable androgens (e.g. 5α-dihydrotestosterone) not only inhibit aromatase activity[33] and stimulate progesterone production in granulosa cells[34] but also induce preantral and antral follicles to undergo atresia[35]. An increased androgen to estrogen ratio has been found in the follicular fluid of atretic follicles and such changes in steroid levels are probably involved in the initiation of atresia[31,36].

Interestingly, the role of androgens as apoptotic factors appears to be tissue specific: on the one hand they induce atresia of follicles and on the other they can act as survival factors in testicular cells[37]. The possible apoptotic or non-apoptotic roles of androgens in other tissues like the endometrium and the breast have not been defined yet and require further investigation.

Higher androgen levels during the middle third of the menstrual cycle are believed to contribute to an increased sex drive around the time of ovulation. Dennerstein and co-workers documented an increase in sexual interest during the follicular and ovulatory phases[38]. Several studies have suggested a positive correlation between sexual desire and the mid-cycle peak in testosterone[39,40]. Nevertheless, due to the many hormonal changes occurring concomitantly around the time of ovulation and in particular because the cyclical changes of estradiol and progesterone can also affect mood and behavior[41], it is difficult in this type of research to relate the changes in sexual interest to a single hormonal factor.

Bancroft and Skakkebaek followed another approach by administering testosterone to women in their fertile period who were sexually unresponsive[42]. The women subsequently showed an increase in the enjoyment of sexual intercourse, sexual thoughts and feelings, frequency of orgasm and vaginal lubrication.

Constant and co-workers presented case reports about two adolescent girls who developed cyclic episodes of psychosis subsequent to menarche[43]. The episodes of extremely violent and sexually provocative behavior coincided with mistimed LH surges (5 days after the onset of menses in case 1 and 5 days prior to menses in case 2). Although androgen levels remained within the normal ranges, the behavioral changes may possibly have been the consequence of variations in serum androgen levels during the cycle. The exacerbations of the psychotic symptoms disappeared when the secretion of gonadotropins was inhibited.

HYPERANDROGENISM AND POLYCYSTIC OVARY SYNDROME

Several conditions have been associated with clinical manifestations of hyperandrogenism (hirsutism, acne and male-pattern baldness). Recent onset and rapid progression of such manifestations raise the suspicion of androgen-secreting neoplasms of the ovary or adrenal glands. Hyperandrogenism may also be the result of Cushing's syndrome or congenital adrenal hyperplasia (CAH; an inherited enzymatic defect). CAH in particular, has been widely studied in relation to cognitive masculinization since affected individuals already produce high levels of androgens prenatally (for a review see ref. 5).

A heterogeneous disorder that more commonly underlies hyperandrogenism is the polycystic ovary syndrome (PCOS). PCOS was originally described by Stein and Leventhal as a triad of hyperandrogenism (hirsutism), anovulation and obesity[44]. A polycystic appearance of the ovaries was originally associated with this disorder. However, this appeared to be secondary to anovulation and is also seen in other anovulatory situations. Furthermore, many women with PCOS do not have multiple ovarian cysts, so that a polycystic aspect is not sufficiently predictive of the disorder. Polycystic ovaries may also be present in about 25% of normally cycling women.

PCOS is estimated to be quite common. The clinical syndrome of hyperandrogenism and oligomenorrhea or amenorrhea may be present in 1–4% of women of reproductive age[45]. About 65% to 85% of all women with androgen excess are diagnosed as having PCOS or as having a variant of PCOS, namely, hyperthecosis[46,47]. The findings in PCOS are variable with obesity in 40–60% of the cases, hirsutism in 60–90%, oligomenorrhea in 50–90%, and infertility in 55–75% of the cases[48]. The exact pathophysiology of PCOS is unknown. Biochemical characteristics that have been considered as being etiologically relevant include elevated serum androgens (free testosterone, androstenedione and DHEA(S)), elevated LH, deviations in luteinizing hormone releasing hormone (LHRH) pulsatility, insulin resistance and low levels of SHBG. Theories about the origin of the syndrome relate to (subtle) enzyme deficiencies in the ovaries and/or adrenals, an exaggeration of normal adrenarche without a specific enzyme defect, endocrinological consequences of obesity (hyperinsulinemia and subsequent stimulation of androgen secretion), elevated estrogens and abnormal gonadotropin secretion[45,49]. Treatment options for the syndrome itself include oral contraceptives, gonadotropin-releasing hormone (GnRH) analogs and antiandrogens, while the resulting infertility can be treated with ovulation-inducing agents.

PCOS provides relevant indications for the consequences of hyperandrogenism in women. Common physical manifestations of (any form of) hyperandrogenism are excess body hair (hirsutism), acne and male-pattern baldness. In this connection, Pepersack and co-workers found that hirsutism in elderly women (without androgen-producing tumors or a history of PCOS) may also indicate higher androgen levels[50]. Increase of muscle development, clitoral hypertrophy and deepening of the voice may also occur as a consequence of hyperandrogenism, but are generally noticed only in cases of marked androgen excess (which is more the case with androgen-producing neoplasms than PCOS). It has been suggested that hyperandrogenism in PCOS contributes to abdominal body fat distribution. However, a study of Pasquali and co-workers indicated that a shift in body fat towards a more masculine distribution may be more the result of hyperinsulinemia associated with PCOS than of hyperandrogenism[51]. Hyperlipidemia may be present in PCOS patients and Geisthövel and co-workers suggested that this may in part be the result of hyperandrogenism[52]. These observations are in fact paradoxical when compared to the effects of androgen therapy in hypogonadal elderly men. An excess of testosterone in women with PCOS will lead to insulin resistance and hyperlipidemia whereas androgen therapy in elderly hypoandrogenic men has been reported to reduce these diabetogenic symptoms[53].

The various relevant manifestations of hyperandrogenism in women, including PCOS, are outlined in Table 2.

EFFECTS OF ANDROGEN ADMINISTRATION IN SURGICALLY AND NATURALLY MENOPAUSAL WOMEN

Surgical menopause

Bilaterally oophorectomized premenopausal women may provide a useful research model for the effects of a reduction in androgen levels, since an important source of androgen production is removed. It has been noticed that in addition to estrogen levels androgen levels also rapidly decline during the first days after surgical removal of the ovaries. Sexual

Table 2 Manifestations of hyperandrogenism in women

Hirsutism
Acne
Oily skin
Oligo- or amenorrhea
Male-pattern baldness
Muscle development
Clitoral hypertrophy
Deepening of the voice
Hyperlipidemia
(Obesity)*

*See text

desire and interest were sometimes found to be lower in oophorectomized women than in non-operated controls, albeit not consistently[54,55]. In this respect, evaluation of the effects of oophorectomy is difficult, since many other factors (partner relationship and social and psychological factors) are involved. For example, some women feel more than others that their femininity is affected by removal of their reproductive organs, and this in itself may have differential effects on sexuality. Furthermore, removal of the ovaries not only results in changed androgen levels, but also in a rapid decline of estrogen levels, which may also affect specific sexual parameters (directly but also indirectly because of the occurrence of vaginal dryness) and behavioral aspects[13–17].

Effects in surgically menopausal women

Another approach is provided by research into the effects of administration of the various sex hormones in surgically menopausal women. A series of well-controlled trials which included androgens was carried out by Barbara Sherwin of McGill University, Montreal (Canada)[14,56–65]. However, when interpreting the results of these studies, it must be kept in mind that the effects noticed may constitute both physiological and pharmacological effects of the sex steroids administered.

Whereas estrogen administration generally results in an alleviation of vaginal dryness and dyspareunia and probably therefore in improvement of sexual satisfaction[56–58,66,67], effects on sexual arousal and desire were generally not seen. In contrast, concomitant testosterone administration increased sexual desire, sexual arousal and the number of fantasies[57–59]. In this respect, androgen–estrogen administration was superior to treatment with estrogen alone and non-treatment or treatment with a placebo. The effects on sexual behavior were moderate or absent, which may be explained by the above-mentioned multifactorial origin of sexuality. Testosterone substitution has also been found to influence physiological aspects of sexual functioning, as could be demonstrated in women who were amenorrheic as a consequence of their lifestyles and had therefore low testosterone levels. Testosterone substitution in combination with erotic stimuli increased vaginal vasocongestion compared to placebo-treated amenorrheic controls exposed to the same erotic film excerpts[68].

Androgen administration in surgically menopausal women may not only affect sexuality, but also other parameters. Short-term testosterone administration in such women resulted in lower depression scores, whether or not given in combination with an estrogen[60]. Hostility scores were higher in women who received androgens, as compared with those receiving estrogens, placebo, or no treatment. The effects disappeared when treatment was stopped. In a similar study, these authors found that androgen–estrogen treatment, or androgen administration alone, resulted in increased energy level, stamina, well-being and appetite and lower somatic and psychological problem scores as compared with estrogen-only or placebo treatment[61]. Superior functioning in the testosterone-treated groups occurred in association with higher plasma testosterone levels during the treatment phases. Long-term androgen–estrogen treatment also resulted in more positive moods, as did long-term estrogen-only treatment, but the combined group felt more composed, elated and energetic than those who were given estrogens alone[62].

The mood responses to androgen–estrogen treatment were found to depend on SHBG levels: higher SHBG levels were associated with less pronounced responses[63]. Although medical castration in men with resulting hypoandrogenism will result in hot flushes in about 70% of recipients[69,70], testosterone administration in women had no effect on flushes, the latter in contrast to estrogens which reduced the flushes[64]. These results were unexpected because testosterone is aromatizable to estrogens and the estrogen levels in the studies were consequently comparable to those observed in non-oophorectomized women. The authors suggested that testosterone administration might have increased SHBG levels (although this suggestion is also controversial)[64]. The higher degree of binding to SHBG would then counteract the effects that were expected from the higher total estrogen levels observed. Another study has nevertheless reported that androgen treatment in women might relieve hot flushes[71].

Administration of androgens, estrogens, or both sex steroids preserved memory and cognitive functioning, which decreased postoperatively in oophorectomized women who received placebo[65]. Finally, it has been suggested that testosterone in women is relevant, together with estrogens, for the preservation of bone mineral density. Watts and co-workers studied bone mineral density in oophorectomized women who were treated with estrogen–androgen or estrogen alone for a duration of 2 years[72]. Spinal bone mineral density significantly increased in the combination treatment group, whereas it did not change in the estrogen-only group. The differences between the two groups were too small to reach statistical significance. A study with ovariectomized rats showed that for the restoration of bone density lower estrogen doses were required when estrogens were supplemented together with non–aromatizable androgens[73]. Consistent with these findings was the observation that estrogens as well as androgens inhibited interleukin-6 production in osteoclast precursor cells and thereby prevented an increase in the production of osteoclasts in bone marrow[74].

In summary, research in oophorectomized women has indicated that testosterone administration may be relevant for sexual desire, arousal and fantasies, mood and well-being (more energy and appetite, and fewer psychological problems and less depression), cognitive function (memory) and possibly preservation of bone mineral density.

Natural menopause and effects of androgen administration

Natural menopause is not associated with a rapid decline in endogenous androgen secretion. Although ovarian function in general declines, the elevated LH levels around the menopause may lead to a (temporarily) increased ovarian secretion of androgens. This probably explains why around the menopause testosterone levels may be reduced, unchanged or increased[75,76]. Thereafter, testosterone and androstenedione may decline to some extent, but not considerably[77]. Only DHEA(S) levels have been documented to decline rapidly with aging, but this process may already start at the age of 25[78].

The effects of testosterone in naturally menopausal women have not been studied as extensively as those in surgically menopausal women.

Nevertheless, some indications are available, although they are sometimes inconsistent.

McCoy and Davidson found that a decline in testosterone levels after the menopause was associated with lower coital frequency[79]. Recently, a greater libido improvement was reported with a combined estrogen–testosterone treatment than with estrogen treatment alone[80]. Myers and co-workers, however, found no increase in sexual behavior or psychophysiologically measured sexual arousal following administration of estrogens or estrogens in combination with testosterone[81]. Both treatments were equally effective in reducing vasomotor symptoms of the climacteric. Raisz and co-workers found that androgen–estrogen treatment and estrogen-only treatment were equally effective with respect to vasomotor symptoms and vaginal dryness, but the combination treatment provided significant relief from psychosomatic and mood-related climacteric complaints, whereas the estrogen-only treatment did not[82]. Both treatments showed a similar decrease in bone resorption markers, but where estrogen treatment alone resulted in a decrease of bone formation parameters, the combined treatment increased these markers[82]. A greater increase in bone density has also been observed in postmenopausal women treated with combined treatment in comparison with estrogen-only treatment[80]. However, in a study with estradiol implants either alone or in combination with a testosterone implant, testosterone conferred no additional bone-sparing effect[83].

Clinical application of testosterone administration to naturally menopausal women, in addition to estrogen–progestin therapy, is controversial[84,85]. Testosterone administration results in an estimated 15–20% of the cases in mild hirsutism. It has been documented that administration of methyltestosterone causes atherogenic lipid profiles[86], whereas non-alkylated androgens[81,87] or aromatizable androgens[88] display more favorable effects. Testosterone–estrogen treatment results in the occurrence of endometrial hyperplasia to a degree similar to estrogen-only treatment, which can be reversed by cyclic progestogen administration[86,89]. Altogether, it is suggested that testosterone administration should only be considered for naturally postmenopausal women who complain about low sexual desire and have relatively low free testosterone levels[84].

The effects of DHEA(S) administration are largely unclear. Epidemiological data indicated that low premenopausal levels of DHEA(S)[90] as well as high postmenopausal levels[91] may contribute to the development of breast cancer. Furthermore, high levels in elderly women were suggested to contribute to insulin resistance, central obesity and cardiovascular disease[92,93]. In such a complex context, only one prospective clinical study that included women has been reported. This small study (17 women of advanced age) indicated that DHEA administration resulted in improvement of physical and psychological well-being, had no side-effects and increased the bioavailability of insulin-like growth factor-I (which was interpreted as a protective effect on aspects of cellular well-being)[94]. However, further studies are needed before final conclusions about the effects of DHEA(S) administration can be drawn.

CONCLUSIONS

Androgens or 'male' hormones are not only relevant for functional aspects of men, but also play distinct roles in women. The evidence for these roles originates from very different sources of research which relate to different phases of the female life cycle (Table 3). If we paste the various pieces of evidence together, the picture emerges that androgens are involved in:

(1) Sexual desire, interest, libido and arousal (but not sexual behavior *per se*);

(2) Cognitive functioning and sex-dimorphic behaviors;

(3) Mood, stamina and well being;

(4) Hirsutism, male-pattern baldness and acne (in case of hyperandrogenism);

(5) Lipid profiles and carbohydrate metabolism;

(6) Bone mineral density; and

(7) Atresia of remaining (non)dominant follicles during the menstrual cycle.

The effects on sexual functioning have been most consistently documented. These latter effects have been noticed during the normal menstrual cycle, to some extent subsequent to surgical removal of the ovaries and subsequent to testosterone administration in surgically menopausal women. In practice, androgen supplementation can be relevant in the treatment of surgically menopausal women (and possibly women with very low androgen levels), as is androgen suppression in women with hyperandrogenism. Future

Table 3 Summary of effects of androgens in women as suggested by the literature

Embryonic development	Female phenotypic sexual differentiation due to lack of androgen
Pre- and perinatal development	Organizational and activational effects on central nervous system influencing gender-typical (cognitive) behaviors
Circulating androgens in later life	Organizational and activational effects on central nervous system influencing gender-typical (cognitive) behaviors
Mid-cycle androgen levels	Selection of the dominant follicle
	Wave of atresia in remaining follicles
	Stimulation of inhibin secretion
	Increased sex drive around the time of ovulation
Hyperandrogenism	Hirsutism, male-pattern baldness and acne
	Infertility
	Hyperlipidemia
Severe hyperandrogenism	Muscle development, clitoral hypertrophy and voice deepening
Surgical menopause (lower androgen levels)	Less sexual desire and interest
	Poorer cognitive functioning
Testosterone administration after surgical menopause	Increased sexual desire, arousal and number of fantasies
	Improvement of mood, energy and well-being
	Preservation of memory and cognitive functioning
	Higher bone mineral density
Testosterone administration after natural menopause	Increased sexual libido
	Higher bone mineral density
Administration of dehydro-epiandrosterone (sulfate) in elderly women	Effects not clear

research will certainly provide us with more supportive and new evidence in reply to that intriguing question: what are the functional roles of androgens in women.

REFERENCES

1. Chang, R. J., Laufer, L. R., Meldrum, D. R., DeFazio, J., Lin, J. H. K. and Vale, W. W. (1983). Steroid secretion in polycystic ovarian disease after

ovarian suppression by a long-acting gonadotropin releasing hormone agonist. *J. Clin. Endocrinol. Metab.*, **56**, 897–903

2. Haning, R. V., Flood, C. A., Hackett, R. J., Loughlin, J. S., McClure, N. and Longcope, C. (1991). Metabolic clearance rate of dehydroepiandrosterone sulfate, its metabolism to testosterone and its intrafollicular metabolism to dehydroepiandrosterone, androstenedione, testosterone and dihydrotestosterone *in vivo*. *J. Clin. Endocrinol. Metab.*, **72**, 1088–95

3. Van Doorn, C., Poortman, J., Thyssen, J. H. H. and Schwarz, F. (1981). Actions and interactions of Δ^5-androstene-3 beta, 17 beta-diol and 17 beta-estradiol in the immature rat uterus. *Endocrinology*, **108**, 1587–93

4. Christiansen, K. and Knussmann, R. (1987). Sex hormones and cognitive functioning in men. *Neuropsychobiology*, **18**, 27–36

5. Berenbaum, S. A., Korman, K. and Leveroni, C. (1995). Early hormones and sex differences in cognitive abilities. *Learning and Individual Differences*, **7**, 303–21

6. Raisman, G. and Field, P. (1971). Sexual dimorphism in the preoptic area of the rat. *Science*, **173**, 731–4

7. Gould, E., Woolley, C. S., Frankfurt, M. and McEwen, B. S. (1990). Gonadal steroids regulate dendrite spine density in hippocampal pyramidal cells in adulthood. *J. Neurosci.*, **10**, 1286–91

8. Nottebohm, F. and Arnold, A. P. (1976). Sexual dimorphism in vocal control areas of the songbird brain. *Science*, **194**, 211–13

9. Arnold, A. P. and Gorski, R. A. (1984). Gonadal steroid induction of structural sex differences in the central nervous system. *Ann. Rev. Neurosci.*, **7**, 413–42

10. Udry, J. R., Morris, N. M. and Kovenock, J. (1995). Androgen effects on women's gendered behaviour. *J. Biosoc. Sci.*, **27**, 359–68

11. Luine, V. N. (1994). Steroid hormone influences on spatial memory. *Ann. NY Acad. Sci.*, **743**, 201–11

12. Watanabe, Y., Gould, E. and McEwen, B. S. (1992). Stress induced atrophy of apical dendrites of hippocampal CA3 pyramidal neurons. *Brain Res.*, **588**, 341–5

13. Halpern, D. F. (1992). *Sex Differences in Cognitive Abilities*. (Hillsdale N J: Laurence Erlbaum Associates)

14. Sherwin, B. B. (1994). Estrogenic effects on memory in women. *Ann. NY Acad. Sci.*, **743**, 213–31

15. Philips, S. M. and Sherwin, B. B. (1992). Effects of estrogen in memory function in surgically menopausal women. *Psychoneuroendocrinology*, **17**, 485–95

16. Paganini-Hill, A. and Henderson, V. W. (1994). Estrogen deficiency and risk of Alzheimer's disease in women. *Am. J. Epidemiol.*, **140**, 256–61

17. Henderson, V. W., Paganini-Hill, A., Emanuel, C. K., Dunn, M. E. and Buckwalter, J. G. (1994). Estrogen replacement therapy in older women. Comparison between Alzheimer's disease cases and nondemented control subjects. *Arch. Neurol.*, **51**, 896–900

18. Gibbs, R. B. (1994). Estrogen and nerve growth factor-related systems in brain. *Ann. NY Acad. Sci.*, **743**, 165–96

19. McEwen, B. S., Gould, E., Orchinik, M., Weiland, N. G. and Woolley, C. S. (1995). Oestrogens and the structural and functional plasticity of neurons: implications for memory ageing and neurodegenerative processes. In: Bock, G. R. and Goode, J. R. (eds.) *Non-reproductive Actions of Sex Steroids* (Ciba Foundation Symposium 191), pp. 52–73. (Chichester: Wiley)

20. Hsieh, Y. L., Hsu, C., Yang, S. L., Hsu, H. K. and Peng, M. T. (1996). Estradiol modulation of neuron loss in the medial division of medial preoptic-nucleus in rats during aging. *Gerontology*, **42**, 18–24

21. Jones, K. J. (1994). Androgenic enhancement of motor neuron regeneration. *Ann. NY Acad. Sci.*, **743**, 141–65

22. Bachevalier, J. and Hagger, C. (1991). Sex differences in the development of learning abilities in primates. *Psychoneuroendocrinology*, **16**, 177–88

23. Williams, C. L. and Meck, W. H. (1991). The organizational effects of gonadal steroids on sexually dimorphic spatial ability. *Psychoneuroendocrinology*, **16**, 155–76

24. Roof, R. L. and Havens, M. D. (1992). Testosterone improves maze performance and induces development of a male hippocampus in females. *Brain Res.*, **572**, 310–13

25. Means, L. W., Alexander, S. R. and O'Neil, M. F. (1992). Those cheating rats: male and female rats use odor trails in water-escape 'working memory' task. *Behav. Neural Biol.*, **58**, 144–51

26. Andrews, J. S. (1996). Possible confounding influence of strain, age and gender on cognitive performance in rats. *Cogn. Brain Res.*, **3**, 251–67

27. Gustafsson, J. A. and Stenberg, A. (1974). Neonatal programming of androgen responsiveness of liver of adult rats. *J. Biol. Chem.*, **249**, 719–23

28. Erickson, G. F. (1993). Normal regulation of ovarian androgen production. *Sem. Reproduct. Endocrinol.*, **11**, 307–12

29. Levine, L. R. (1993). Menstrual physiology. In Brown, J. S. and Crombleholme, W. R. (eds.) *Handbook of Gynecology and Obstetrics*, pp. 12–19. (Englewood Cliffs: Prentice Hall)

30. Judd, H. L. and Yen, S. S. C. (1973). Serum androstenedione and testosterone levels during the menstrual cycle. *J. Clin. Endocrinol. Metab.*, **36**, 475–81

31. Hsueh, A. J. W., Billig, H. and Tsafriri, A. (1994). Ovarian follicle atresia: a hormonally controlled apoptotic process. *Endocr. Rev.*, **15**, 707–24

32. Haning, R. V., Hackett, R. J., Flood, C. A., Loughlin, J. S., Zhao, Q. Y. and Longcope, C. (1993). Testosterone, a follicular regulator: key to anovulation. *J. Clin. Endocrinol. Metab.*, **77**, 710–15

33. Hillier, S. G., Van der Boogaard, A. M. Y., Reichert, L. E. J. and Van Hall, E. V. (1980). Alterations in granulosa cell aromatase activity accompanying preovulatory follicular development in the rat ovary with evidence that 5α-reductase C19 steroids inhibit the aromatase reaction *in vitro*. *J. Endocrinol.*, **84**, 409–19

34. Schomberg, D. W., Stouffer, R. L. and Tyrey, I. (1976). Modulation of progesterone secretion in ovarian cells by 17 beta-hydroxy-5 alpha-androstan-3-one (dihydrotestosterone): a direct demonstration in monolayer culture. *Biochem. Biophys. Res. Commun.*, **68**, 77–81

35. Azzolin, G. C. and Saiduddin, S. (1983). Effect of androgens on the ovarian morphology of the hypophysectomized rat. *Proc. Soc. Exp. Biol. Med.*, **172**, 70–3

36. Carson, R. S., Findlay, J. K., Clarke, I. J. and Burger, H. B. (1981). Estradiol, testosterone and androstenedione in ovine follicular fluid during growth and atresia of ovarian follicles. *Biol. Reprod.*, **24**, 105–13

37. Tapanainen, J. S., Tilly, J. L., Vihko, K. K. and Hsueh, A. J. W. (1993). Hormonal control of apoptotic cell death in the testis: gonadotropins and androgens as testicular cell survival factors. *Mol. Endocrinol.*, **7**, 643–50

38. Dennerstein, L., Gotts, G., Brown, J. B., Morse, C. A., Farley, T. M. M. and Pinol, A. (1994). The relationship between the menstrual cycle and female sexual interest in women with PMS complaints and volunteers. *Psychoneuroendocrinology*, **19**, 293–304

39. Bancroft, J., Sanders, D., Davidson, D. and Warner, P. (1983). Mood, sexuality, hormones, and the menstrual cycle. III. Sexuality and the role of androgens. *Psychosom. Med.*, **45**, 509–16

40. Abplanalp, J. M., Rose, R. M., Donnelly, A. F. and Livingston-Vaughan, L. (1979). Psychoendocrinology of the menstrual cycle: II. The relationship between enjoyment of activities, moods and reproductive hormones. *Psychosom. Med.*, **41**, 605–15

41. Bäckström, T. (1995). Symptoms related to the menopause and sex steroid treatments. In Bock, G. R. and Goode, J. R. (eds.) *Non-reproductive Actions of Sex Steroid* (Ciba Foundation Symposium 191) pp. 171–86. (Chichester: Wiley)

42. Bancroft, J. and Skakkebaek, N. E. (1979). Androgens and human sexual behaviour. In *Sex Hormones and Behaviour* (Ciba Foundation Symposium 62 new series) pp. 209–26. (Amsterdam: Excerpta Medica)

43. Constant, M., Abrams, C. A. L. and Chasalow, F. L. (1993). Gonadotropin-associated psychosis in premenstrual behaviour disorder. *Horm. Res.*, **40**, 141–4

44. Stein, I. F. and Leventhal, M. L. (1935). Amenorrhoea associated with bilateral polycystic ovaries. *Am. J. Obstet. Gynecol.*, **29**, 181

45. Rittmaster, R. S. (1993). Hyperandrogenism. In Copeland, L. J. (ed.) *Textbook of Gynecology* pp. 414–46. (Philadelphia: Saunders)

46. Maroulis, G. B. (1981). Evaluation of hirsutism and hyperandrogenemia. *Fertil. Steril.*, **36**, 273–305

47. Rittmaster, R. S. and Loriaux, D. L. (1987). Hirsutism. *Ann. Intern. Med.*, **106**, 95–107

48. American Society for Reproductive Medicine (Formerly The American Fertility Society). (1995). *The Evaluation and Treatment of Androgen Excess; Guidelines for Practice.* (Birmingham (AL): American Society for Reproductive Medicine)

49. Guzick, D. S., Wing, R., Smith, D., Berga, S. L. and Winters, S. J. (1994). Endocrine consequences of weight loss in obese, hyperandrogenic, anovulatory women. *Fertil. Steril.*, **61**, 589–604

50. Pepersack, T., Rossi, C., Dupuis, F., Lefevre, A., Vanhaeverbeek, M. and Dekoninck, W. (1993). Hormonal status and clinical relevance of hirsutism in elderly women. *Acta Endocrinol.*, **129**, 307–10

51. Pasquali, R., Casimirri, F., Cantobelli, S., Labate, A. M. M., Venturoli, S., Paradisi, R. and Zannarini, L. (1993). Insulin and androgen relationships with abdominal body fat distribution in women with and without hyperandrogenism. *Horm. Res.*, **39**, 179–87

52. Geisthövel, F., Olbrich, M., Frorath, B., Thiemann, M. and Weitzell, R. (1994). Obesity and hypertestosteronaemia are independently and synergistically associated with elevated insulin concentrations and dyslipidaemia in pre-menopausal women. *Hum. Reprod.*, **9**, 610–16

53. Haffner, S. M. (1996). Androgens in relation to cardiovascular disease and insulin resistance in aging men. In Oddens, B. J. and Vermeulen, A. *Androgens and the Aging Male*, pp. 65–83. (Carnforth: Parthenon)

54. Hawton, K., Gath, D. and Day, A. (1994). Sexual function in a community sample of middle-aged women with partners: effects of age, socioeconomic, psychiatric, gynecological, and menopausal factors. *Arch. Sex. Behav.*, **23**, 375–95

55. Dennerstein, L., Smith, A. M. A., Morse, C. A. and Burger, H. (1994). Sexuality and the menopause. *J. Psychosom. Obstet. Gynecol.*, **15**, 59–66

56. Sherwin, B. B. (1991). The impact of different doses of estrogen and progestin on mood and sexual behaviour in postmenopausal women. *J. Clin. Endocrinol. Metab.*, **72**, 336–43

57. Sherwin, B. B., Gelfand, M. M. and Brender, W. (1985). Androgen enhances sexual motivation in females: a prospective, crossover study of sex steroid administration in the surgical menopause. *Psychosom. Med.*, **47**, 339–51

58. Sherwin, B. B. and Gelfand, M. M. (1987). The role of androgen in the maintenance of sexual functioning in oophorectomized women. *Psychosom. Med.*, **49**, 397–409

59. Sherwin, B. B. (1985). Changes in sexual behaviour as a function of plasma sex steroid levels in postmenopausal women. *Maturitas*, **7**, 225–33

60. Sherwin, B. B. and Gelfand, M. M. (1985). Sex steroids and effect in the surgical menopause: a double-blind, cross-over study. *Psychoneuroendocrinology*, **10**, 325–35

61. Sherwin, B. B. and Gelfand, M. M. (1985). Differential symptom response to parenteral estrogen and/or androgen administration in the surgical menopause. *Am. J. Obstet. Gynecol.*, **151**, 153–60

62. Sherwin, B. B. (1988). Affective changes with estrogen and androgen replacement therapy in surgically menopausal women. *J. Affective Disord.*, **14**, 177–87

63. Sherwin, B. B. and Gelfand, M. M. (1984). Effects of parenteral administration of estrogen and androgen on plasma hormone levels and hot flushes in the surgical menopause. *Am. J. Obstet. Gynecol.*, **148**, 552–7

64. Sherwin, B. B. and Gelfand, M. M. (1987). Individual differences in mood with menopausal replacement therapy: possible role of sex hormone binding globulin. *J. Psychosom. Obstet. Gynaecol.*, **6**, 121–31

65. Sherwin, B. B. (1988). Estrogen and/or androgen replacement therapy and cognitive functioning in surgically menopausal women. *Psychoneuroendocrinology*, **13**, 345–57

66. Maoz, B. and Durst, N. (1980). The effects of oestrogen therapy on the sex life of post-menopausal women. *Maturitas*, **2**, 327–36

67. Nathorst-Boos, J., Wiklund, I., Matsson, L. A., Sandin, K. and Von Schoultz, B. (1993). Is sexual life influenced by transdermal estrogen therapy? *Acta Obstet. Gynecol. Scand.*, **72**, 656–60

68. Tuiten, A., Laan, E., Panhuysen, G., Everaerd, W., de Haan, E., Koppeschaar, H. and Vroon, P. (1996). Discrepancies between genital responses and subjective sexual functioning during testosterone substitution in women with hypothalamic amenorrhea. *Psychosom. Med.*, **58**, 234–41

69. Karling, P., Hammar, M. and Varenhorst, E. (1994). Prevalence and duration of hot flushes after surgical or medical castration in men with prostatic carcinoma. *J. Urol.*, **152**, 1170–3

70. Levy, A. (1996). Why flush? *Lancet*, **347**, 73–4

71. Greenblatt, R. B. (1987). Is there a place for androgens in gynecological disorders? *Gynecol. Endocrinol.*, **1**, 209–19

72. Watts, N. B., Notelovitz, M., Timmons, M. C., Addison, W. A., Wiita, B. and Downey, L. J. (1995). Comparison of oral estrogens and estrogens plus androgen on bone mineral density, menopausal symptoms, and lipid–lipoprotein profiles in surgical menopause. *Obstet. Gynecol.*, **85**, 529–37

73. Ederveen, A. G. H. and Kloosterboer, H. J. (1995). A non-aromatisable androgen potentiates the effect of an estrogen on the prevention of ovariectomy-induced bone loss in the rat. *Calcif. Tissue Int.*, **56**, 484

74. Manolagas, S., Bellido, T. and Jilka, R. (1995). Sex steroids, cytokines and the bone marrow: new concepts on the pathogenesis of osteoporosis. In Bock, G. R. and Goode, J. R. (eds.) *Non-reproductive Actions of Sex Steroids* (Ciba Foundation Symposium 191) pp. 187–202. (Chichester: Wiley)

75. Longcope, C., Hui, S. L. and Johnston, C. C. (1987). Free estradiol, free testosterone and sex hormone-binding globulin in perimenopausal women. *J. Clin. Endocrinol. Metab.*, **64**, 513–18

76. Longcope, C., Franz, C., Morello, C., Baker, R. and Johnston, C. C. (1986). Steroid and gonadotropin levels in perimenopausal women during the perimenopausal years. *Maturitas*, **8**, 189–96

77. Rannevik, G., Jeppsson, S., Bjerre, B., Laurell-Borulf, Y. and Svanberg, L. (1995). A longitudinal study of the perimenopausal transition: altered profiles of steroid and pituitary hormones, SHBG and bone mineral density. *Maturitas*, **21**, 103–13

78. Carlström, K., Brody, S., Lunell, N. O., Lagrelius, A., Möllerström, G., Pousette, A., Rannevik, G., Stege, R. and Von Schoultz, B. (1988).

Dehydroepiandrosterone sulphate and dehydroepiandrosterone in serum: differences related to age and sex. *Maturitas*, **10**, 297–306

79. McCoy, N. L. and Davidson, J. M. (1985). A longitudinal study of the effects of menopause on sexuality. *Maturitas*, **7**, 203–10

80. Davis, S. R., McCloud, P., Strauss, B. J. G. and Burger, H. (1995). Testosterone enhances estradiol's effects on postmenopausal bone density and sexuality. *Maturitas*, **21**, 227–36

81. Myers, L. S., Dixen, J., Morrissette, D., Carmichael, M. and Davidson, J. M. (1990). Effects of estrogen, androgen and progestin on sexual psychophysiology and behaviour in postmenopausal women. *J. Clin. Endocrinol. Metab.*, **70**, 1124–31

82. Raisz, L. G., Wiita, B., Artis, A., Bowen, A., Schwartz, S., Trahiotis, M., Shoukri, K. and Smith, J. (1996). Comparison of the effects of estrogen alone and estrogen plus androgen on biochemical markers of bone formation and resorption in postmenopausal women. *J. Clin. Endocrinol. Metab.*, **81**, 37–43

83. Garnett, T., Studd, J., Watson, N., Savons, M. and Leather, A. (1992). The effects of plasma estradiol levels on increases in vertebral and femoral bone density following therapy with estradiol and testosterone implants. *Obstet. Gynecol.*, **79**, 968–72

84. Gelfand, M. M. (1994). Treating menopausal symptoms with estrogen and androgen preparations. *IM Int. Med.*, **15**, 51–3

85. Sands, R. and Studd, J. (1995). Exogenous androgens in postmenopausal women. *Am. J. Med.*, **98**, (Suppl 1A), 76–9

86. Hickok, L. R., Toomey, C. and Speroff, L. (1993). A comparison of esterified estrogens with and without methyltestosterone: effects on endometrial histology and serum lipoproteins in postmenopausal women. *Obstet. Gynecol.*, **82**, 919–24

87. Sherwin, B. B., Gelfand, M. M., Schucher, R. and Gabor, J. (1987). Postmenopausal estrogen and androgen replacement and lipoprotein lipid concentrations. *Am. J. Obstet. Gynecol.*, **156**, 414–19

88. Bagatell, C. J. and Bremner, W. J. (1995). Androgen and progestagen effects on plasma lipids. *Progr. Cardiovasc. Dis.*, **38**, 255–71

89. Gelfand, M. M., Ferency, A. and Bergeron, C. (1989). Endometrial response to estrogen–androgen stimulation. *Prog. Clin. Biol. Res.*, **320**, 29–40

90. Helzlsouer, K. J., Gordon, G. B., Alberg, A., Bush, T. L. and Comstock, G. W. (1992). Relationship of prediagnostic serum levels of DHEA and DS to the risk of developing premenopausal breast cancer. *Cancer Res.*, **52**, 1–4

91. Gordon, G. B., Bush, T. L., Helzlsouer, K. J., Miller, S. R. and Comstock, G. W. (1990). Relationship of serum levels of dehydroepiandrosterone and dehydroepiandrosterone sulphate to the risk of developing postmenopausal breast cancer. *Cancer Res.*, **50**, 3859–62

92. Barrett-Connor, E. and Khaw, K. T. (1987). Absence of an inverse relation of dehydroepiandrosterone sulphate with cardiovascular mortality in post-menopausal women. *N. Engl. J. Med.*, **317**, 711

93. Ebeling, P. and Koivisto, V. A. (1994). Physiological importance of dehydroepiandrosterone. *Lancet*, **343**, 1479–81

94. Morales, A. J., Nolan, J. J., Nelson, J. C. and Yen, S. S. C. (1994). Effects of replacement dose of dehydroepiandrosterone in men and women of advancing age. *J. Clin. Endocrinol. Metab.*, **78**, 1360–7

DISCUSSION

Schiavi: We have studied in our laboratories the role of testosterone during the normal menstrual cycle in a large group of women, and looked at sexual desire and arousability. The latter was measured by subjective reporting following exposure to erotic tapes as well as by vaginal plethysmography. Actually, we found a relationship between circulating androgens and those two dimensions, within the normal variations of the menstrual cycle. A relevant point to consider in this research field is the fact that enhancing sexual desire does not necessarily imply enhancement of sexual satisfaction. In some instances, in particular where there is a lack of congruence between the man and woman, increases in the sexual desire of the woman might be quite disruptive for the marital relationship. I think that sexual satisfaction is a neglected parameter that needs to be taken into account in this type of research.

Bergink: To what extent is progesterone important to sexual desire? I notice that the desire may be quite low during the luteal phase of the menstrual cycle.

Schiavi: We have not found any relation between progesterone and variations in sexual desire within the normal menstrual cycle.

Longcope: With reference to the potential enhancing effects of androgens on sexual desire in women, there is evidence that androgens do enhance motivation[1]. This may help to explain the report that oral contraceptives decrease sexual desire at ovulation[2], since the oral contraceptives decrease androgen levels.

Bergink: Free testosterone levels are decreased in a very consistent fashion during the use of oral contraceptives. Therefore, the effect of testosterone on sexual desire is expected to be less. This seems to be paradoxical because I am not aware that such a consistent decrease has been observed during the use of oral contraceptives. Other factors may, therefore, also seem to play a role.

Ginsburg: I wonder whether the progestogen component may provide an explanation. My experiences with the effects of the oral contraceptive pill are limited, but with menopausal therapy, I noticed that the progestogen used may have a profound influence on the woman's libido, arousal and vaginal lubrication. The effects of the progestogens in the oral contraceptive pill may be relevant in this connection.

Thijssen: With respect to the effects of androgens on sexuality, I recall having read a recently published double-blind controlled study, in which effects of tibolone on sexual performance, interest, etc. were studied[3]. Tibolone is a form of hormone replacement therapy with, among other things, estrogenic and some androgenic properties. After six months of treatment tibolone significantly increased sexual arousal, interest, frequency and so on.

Ginsburg: I can confirm that we have had a few women who actually stopped taking tibolone because of increased sexual arousal. They either had no partner, a partner without sexual interest, or who was impotent. Therefore, in these circumstances, an increase in sexual desire was an adverse effect.

REFERENCES

1. Sherwin, B. B., Gelfand, M. M. and Brender, W. (1985). Androgen enhances sexual motivation in females: a prospective, crossover study of sex steroid administration in the surgical menopause. *Psychosom. Med.*, **47**, 339–51

2. Adams, D. B., Gold, A. R. and Burt, A. D. (1978). Rise in female-initiated sexual activity at ovulation and its suppression by oral contraceptives. *N. Engl. J. Med.*, **299**, 1145–50

3. Palacios, S., Menendez, C., Jurado, A. R., Castano, R. and Vargas, J. C. (1995). Changes in sex behaviour after menopause: effects of tibolone. *Maturitas*, **22**, 155–63

11

Effects of androgen supplementation in the aging male

J. L. Tenover

A decline in the serum level of testosterone accompanies normal aging in men, but the physiological and clinical impacts of this decline are still in question. Unlike females who uniformly become hypogonadal with menopause, not all men will develop classic hypogonadal androgen levels. It may be, however, that many older men are relatively androgen deficient and could benefit from androgen supplementation. Since there is no single measurement of clinically significant 'hypogonadism', many older men who have serum testosterone levels near or below the normal range established in young adult men might show physiological gains in androgen target organs with hormone replacement. The purpose of this brief review is to highlight what is known about androgen replacement therapy in older men.

Although androgens have important physiological actions in a large number of tissues, their effects in the older male have focused on those listed in Table 1. Age-related decrements in muscle, bone and the central nervous system are a part of normal aging, and testosterone (and other androgens) are under evaluation as potential 'trophic factors' that might either slow or partially reverse some of these decrements. Of course, any discussion of the use of androgens as a possible trophic factor to prevent frailty or disability in the aging male must also consider the possible risks of testosterone supplementation in this age group. Possible detrimental effects of androgens would include:

(1) Fluid retention resulting in elevation in blood pressure, edema, or exacerbation of congestive heart failure;

(2) Effects on erythropoiesis (development of polycythemia);

(3) Exacerbation of sleep apnea;

(4) Changes in the cardiovascular system through effects on lipoproteins, hemostasis, or through other mechanisms; and

(5) Effects on the prostate, resulting in promotion of benign hyperplasia or pre-existent prostate cancer.

ANDROGENS AND BODY COMPOSITION

Since normal aging results in an increase in upper and central body fat, a decrease in muscle mass and a decline in some aspects of muscle strength, androgens, long noted for their anabolic effects, have begun to be evaluated as a modality for improving body composition and strength in the elderly. In one study[1] 13 men, mean age 67.5 years, with serum testosterone levels equal to or below 400 ng/dl (13.9 nmol/l), were treated in a double-blind, placebo-controlled, cross-over study with testosterone enanthate (100 mg/week) or placebo for 3 months each. Body composition was measured by hydrostatic weighing. At the end of the testosterone treatment phase, there was an increase in body weight (an average increase of 1.5 kg over that during placebo treatment or baseline), an increase in lean body mass (1.8 kg over that during placebo treatment or baseline), and a tendency for a decrease in total body fat (27.9% during testosterone treatment compared to 28.8% at baseline or 28.9% during placebo therapy). Body circumference measurements and grip strength were not noted to change with androgen therapy in this study.

Another study[2] involved 14 men, ages 69–89 years, with 'bioavailable' testosterone less than 70 ng/dl. Eight of these men received testosterone enanthate (200 mg every 2 weeks) for 3 months, and six men received no treatment. In this study, handgrip strength increased with the androgen therapy compared to the non-treatment group. These

Table 1 Target areas of interest for androgen therapy in aging men

Body composition	Sleep
muscle	Prostate
fat	Erythropoiesis
strength	Cardiovascular
Bone	lipoproteins
Sexual function	hemostasis
Cognition/Mood	Fluid volume

results were confirmed by the same group in a larger cohort of men ($n = 31$)[3]. Another group has reported that in eight men with baseline serum testosterone levels of less than 250 ng/dl, some of whom were in the older age range, testosterone therapy for 6 months with the scrotal patch resulted in a 6% decrease in body fat and a 5% increase in lean body mass, as determined by dual energy X-ray absorptiometry[4]. Strength, as measured by dynamometry, also increased with testosterone therapy.

A study in obese men over the age of 40 (mean age 56.7 years) who were treated for 9 months with a daily application of a gel preparation of testosterone applied to the arms and shoulders, demonstrated a 9% decline in visceral adipose tissue, as measured by computerized tomography (CT) scan without a change in overall lean body mass[5]. Katznelson and co-workers have reported that in men up to age 69 years (mean age of study participants was 59 years), 18 months of testosterone ester therapy at 100 mg/week resulted in a 20% decline in body fat, and a 5% increase in lean body mass[6]. A summary of the results of testosterone therapy on body composition in older men is presented in Table 2.

Although testosterone replacement studies in older men have varied in design, and mode and duration of therapy, the data consistently demonstrate changes in body composition with treatment – a decrease in body fat, increase in lean body mass, and perhaps a change in some aspects of muscle

Table 2 Summary of body composition results for studies involving testosterone therapy to older men

Study (Ref.)	Age (years)	Length of treatment (months)	Treatment type	Fat mass	Lean body mass	Strength
1	67.5 (mean)	3	TE*	NC	↑	NC
2	69–89	3	TE	—	—	↑
3	50–90	6	TE	—	—	↑
4	various	6	S Patch†	↓	↑	↑
5	56.7 (mean)	9	Gel prep	↓	NC	—
6	59 (mean)	18	TE	↓	↑	—

*TE, testosterone ester; †S Patch, scrotal patch; NC, no change; ↑, increase; ↓, decrease

strength. It will be important, however, for larger studies to corroborate these data, to determine if effects are sustained, and to elucidate if the increase in muscle strength with testosterone will be of such magnitude as to have a clinically meaningful impact on function. There are a number of studies of testosterone therapy in older men currently in progress that may add some information in this regard.

ANDROGENS AND BONE

As long ago as 1948, Albright and Reifenstein reported that testosterone given to a 72-year-old man with osteoporosis resulted in a decline in total calcium excretion[7]. Lafferty and colleagues, in 1964, reported similar results in a 75-year-old osteopenic male[8]. Oppenheim and Klibanski gave testosterone for 6–8 months to six older hypogonadal (mean testosterone = 144 ng/dl) men (mean age 61 years) with spinal bone densities below those of age-matched controls[9]. Spinal bone densities were reported to increase in all six men.

Several recent studies have evaluated the effect of testosterone on parameters of bone turnover and/or bone density in older men. In two studies previously cited[1,2] a significant decline in urinary excretion of hydroxyproline, or an increase in serum osteocalcin, were reported in men after 3 months of testosterone therapy. After 18 months of testosterone therapy in 12 older men, another study noted a 5–13% increase in trabecular bone density, depending on modality of measurement used[6]. The same group using 6 months of testosterone enanthate at 100 mg/week noted a decrease in both urinary pyridinoline cross-links and serum osteocalcin, suggesting a decline in bone turnover[10]. Finally a study in four hypogonadal men (ages 38–74 years), given 1 year of testosterone enanthate therapy (200 mg every 2 weeks), reported a 23% increase in lumbar bone mineral density at one year[11]. These data are summarized in Table 3.

Overall, there are studies that demonstrate that bone density may be increased and bone turnover slowed in elderly men on testosterone therapy, although additional longer term data are needed to confirm that androgen replacement can sustain a stabilization or reversal of bone loss in this age group.

Table 3 Testosterone therapy effects on bone in older men

Study (Ref.)	Length of treatment (months)	Study n	Parameters of bone turnover		Bone density (Lumbar spine)
			Osteocalcin	Hydroxyproline/ pyridinolines	
1	3	13	NC	↓	—
2	3	8	↑	—	—
6	18	12	—	—	↑
9	6–8	6	—	—	↑
10	6	30	↓	↓	—
11	12·	4	—	—	↑

↓, decrease; ↑, increase; NC, no change

ANDROGENS AND MOOD, COGNITION AND SEXUAL BEHAVIOR

In elderly men, the effects of testosterone therapy on mood, aspects of cognition and sexual behavior have not been extensively evaluated. Several studies have shown improvement in spatial cognition with androgen therapy[12,13], while others have reported improvement in sense of well-being[1,5,14] and/or an increase in libido[14,15] with testosterone. Impotence in older men is usually multifactorial and testosterone therapy often has not been beneficial.

THE POTENTIAL RISKS OF ANDROGEN THERAPY, ESPECIALLY IN OLDER MEN

Androgen supplementation carries some potential risks, especially in the older man who may have certain co-existent medical problems. Among the specific risks that need to be considered are water retention, development of polycythemia, hepatotoxicity, exacerbation of sleep apnea, development of detrimental effects on the cardiovascular system and exacerbation of pre-existent benign or malignant prostate disease.

Water retention is known to occur with androgen therapy, and it is possible that testosterone supplementation in older men could lead to hypertension, peripheral edema, or exacerbation of congestive heart failure. In the four studies of androgen therapy in older men that have reported on fluid status[1,2,5,16] a total of 41 men received androgens from 3 to 9 months and no man was reported to have developed hypertension, edema, or congestive heart failure. However, some of the increased weight noted in two of the studies[1,2] was thought to have been due to water retention.

The stimulatory effects of androgens on erythropoiesis are well documented. Studies in which young hypogonadal or eugonadal men have been supplemented with testosterone have shown an increase in hemoglobin and hematocrit levels with treatment. Similar results have been reported in the studies of androgen therapy in older men (Table 4). Because healthy older men tend to have slightly lower hematocrits and hemoglobin levels than do young adult men[17], except in those men with other clinical causes that could elevate hemoglobin levels, the erythropoietic effects of testosterone in the older age group may not uniformly lead to problems with polycythemia. Furthermore, the method and extent of testosterone replacement may impact the effects on erythropoiesis, with presumably less of an effect being seen with lower, more frequent doses of testosterone esters or with the scrotal or transdermal patch.

Although hepatotoxicity has been reported with administration of some oral androgens (notably methyl-testosterone), it is not usually seen with non-oral forms of testosterone supplementation in young adult men. There have been no reports of hepatotoxicity (abnormal liver enzyme blood tests) in any of the studies of testosterone therapy in older men.

Testosterone administration has been reported to exacerbate pre-existing sleep apnea, but none of the reports of testosterone therapy in older men have noted the development of sleep apnea. However, several of the studies have prescreened study participants for sleep apnea, a procedure that would

Table 4 Hematocrit changes in elderly men with testosterone ester (testosterone enanthate or cypionate) supplementation

Study (Ref.)	Treatment dose	Treatment duration (months)	Mean change in Hct* (%)	Developing Hct >51%
1	100 mg/week	3	+3.6 (8%)	2/13
2	200 mg/2 week	3	+7.0 (17%)	ND
14	25–50 mg/2 week	24	no change	none
16	200 mg/2 week	3	+4.7 (ND)	1/11
18	150 mg/70kg/2 week	6	+ (ND)	2/8
20	300 mg/3 week	3	+ (19%)	9/15

Hct, hematocrit; ND, no data

seem prudent if androgen therapy is considered. In addition, data from one study led the author to suggest that the sleep apnea syndrome may be one mechanism for development of polycythemia seen with testosterone replacement[18].

A portion of the increase in mortality attributed to cardiovascular disease in men compared with premenopausal women may be the result of androgens, although it has also been suggested that low androgen levels in men may contribute to the development of cardiovascular disease at older age. Some of the increased cardiovascular risk in men as compared with women may be mediated through androgen effects on serum lipoprotein levels or through other factors such as modulation of plasma levels of the vasoconstrictor endothelin or effects on hemostasis.

The data on effects of testosterone therapy on serum cholesterol levels in older men are summarized in Table 5. In general, parenteral testosterone therapy leads to a decrease in total and low-density lipoprotein-cholesterol levels, with no change in high-density lipoprotein-cholesterol levels. These changes in cholesterol levels with testosterone therapy have been modest and the ultimate impact on cardiovascular risk is unclear. Effects of testosterone therapy on other factors affecting cardiovascular risk have not been reported in the elderly.

The role of androgen supplementation as a potentiator of benign or malignant prostate disease in the older male is based largely on theoretical grounds at this point, but it is one of the areas of highest concern in regard to the long-term safety of testosterone therapy. Many of the studies of testosterone replacement in older men have monitored the effect of therapy on prostate-specific antigen (PSA) levels, prostate size, and/or prostatic symptoms. Of six testosterone studies in older men that have monitored for

Table 5   Effect of testosterone therapy on serum cholesterol levels in older men

Study (Ref.)	n	Treatment Duration (months)	Average percent change				
				Total Chol	HDL-Chol		LDL-Chol
5	20	9	↓	11.7%	NC		—
2	8	3	↓	9%	NC		—
1	13	3	↓	11.0%	NC	↓	11.2%
14	10	12	↓	*	NC	↓	*

Chol, cholesterol; ↓, decrease; NC, no change; *, exact change not stated prostate changes[1,2,3,5,14,19], only one has noted any significant change in prostate parameters with therapy, a 21% increase in serum PSA from baseline values[1]. However, all these studies were of relatively short duration (3–24 months), which could limit the ability to detect any augmentation of benign or malignant prostate growth. Clearly, further data with regard to the long-term effects of testosterone therapy on the prostate in older men need to be generated before the relative risk of such therapy can be fully evaluated.

CONCLUSION

A number of studies, involving small numbers of subjects, suggest that older men, with serum testosterone levels that are near or below the lower end of the normal range for young adult men, may be a group in whom testosterone replacement therapy might benefit bone, muscle and psychosexual functions. These studies have also demonstrated that, in the short term, and with adequate prescreening, significant adverse effects seem to have been averted. However, results from longer term and larger clinical studies are needed before a risk/benefit profile for testosterone therapy in 'hypogonadal' older men can be assessed. At the current time, large scale use of testosterone replacement therapy in the older man would be premature.

REFERENCES

1. Tenover, J. S. (1992). Effects of testosterone supplementation in the aging male. *J. Clin. Endocrinol. Metab.*, **75**, 1092–8
2. Morley, J. E., Perry, H. M., Kaiser, F. E., Kraenzle, D., Jensen, J., Houston, K., Mattammal, M. and Perry, H. M. (1993). Effects of testosterone replacement therapy in old hypogonadal males: a preliminary study. *J. Am. Geriatr. Soc.*, **41**, 149–52
3. Sih, R., Perry, H. M., Kaiser, F. E., Patrick, P., Ross, C. and Morley, J. E. (1995). Testosterone therapy increases strength in older hypogonadal men. *J. Invest. Med.*, **43(2)** Suppl. 2, 300A
4. Haddad, G., Peachey, H., Slipman, C. and Snyder, P. J. (1994). Testosterone treatment improves body composition and muscle strength in hypogonadal men. In *Program and Abstracts of the 76th Annual Meeting of the Endocrine Society* (USA), Abstract #1302, p. 506

5. Marin, P., Holmang, S., Gustafsson, C., Jonsson, L., Kvist, H., Elander, A., Eldh, J., Sjostrom, L., Holm, G. and Bjorntorp, P. (1993). Androgen treatment of abdominally obese men. *Obesity Res.*, **1**, 245–51

6. Katznelson, L., Finkelstein, J., Baressi, C. and Klibanski, A. (1994). Increase in trabecular bone density and altered body composition in androgen replaced hypogonadal men. In *Program and Abstracts of the 76th Annual Meeting of the Endocrine Society* (USA), Abstract #1524, p. 581

7. Albright, F. and Reifenstein, E. C. (1948). Metabolic bone disease: osteoporosis. In Albright, F. and Reifenstein, E. C. (eds.) *The Parathyroid Glands and Metabolic Bone Disease*, pp. 145–332. (Baltimore: Williams and Wilkins)

8. Lafferty, F. W., Spencer, G. E. and Pearson, O. H. (1964). Effects of androgens, estrogens, and high calcium intakes on bone formation and resorption in osteoporosis. *Am. J. Med.*, **36**, 514–28

9. Oppenheim, D. and Klibanski, A. (1989). Osteopenia in men with acquired hypogonadism: improvement with testosterone therapy. In *Program and Abstracts of the 71st Annual Meeting of the Endocrine Society* (USA), Abstract #585, p. 289

10. Katznelson, L., Finkelstein, J. S. and Klibanski, A. (1995). Effect of androgen replacement therapy on bone turnover in adult men with acquired hypogonadism. In *Program and Abstracts of the 77th Annual Meeting of the Endocrine Society* (USA), Abstract #P2–123, p. 321

11. Tomasic, P. V., Sollock, R. L., Armstrong, D. W. and Shakir, K. M. M. (1994). Osteoporosis in men with borderline idiopathic hypogonadotropic hypogonadism. In *Program and Abstracts of the 76th Annual Meeting of the Endocrine Society* (USA), Abstract #1043, p. 461

12. Janowsky, J. S., Oviatt, S. K. and Orwoll, E. S. (1994). Testosterone influences spatial cognition in older men. *Behav. Neurosci.*, **108(2)**, 325–32

13. Orwoll, E., Oviatt, S. and Biddle, J. (1992). Transdermal testosterone supplementation in normal older men. In *Program and Abstracts of the 74th Annual Meeting of the Endocrine Society* (USA), Abstract #1071, p. 319

14. Ellyin, F. M. (1995). The long-term beneficial effects of low dose testosterone in the aging male. In *Program and Abstracts of the 77th Annual Meeting of the Endocrine Society* (USA), Abstract #P2–127, p. 322

15. Tenover, J. S. (1992). Effect of testosterone (T) and 5a-reductase inhibitor (5-ARI) administration on responses to a sexual function questionnaire in older men. In *Program and Abstracts of the 17th Annual Meeting of the American Society of Andrology*, Abstract #129, p. 50

16. Sih, R., Kaiser, F. E., Morley, J. E., Sawardekar, M. A., Patrick, P. and Perry, H. M. (1994). Testosterone increases strength in older hypogonadal men. In *Program and Abstracts of the 31st American Geriatrics Society Meeting*, Abstract #A25, p. SA7

17. Williamson, C. S. (1916). Influence of age and sex on hemoglobin: a spectro-photometric analysis of nine hundred and nineteen cases. *Arch. Intern. Med.*, **18**, 505–12

18. Drinka, P. J., Jochen, A. L., Cuisinier, M., Bloom, R., Rudman, I. and Rudman, D. (1995). Polycythemia as a complication of testosterone replacement therapy in nursing home men with low testosterone levels. *J. Am. Geriatr. Soc.*, **43**, 899–901

19. Atkinson, L., Cook, D., Cunningham, G., Grant, M. and Lin, T. (1995). Chronic replacement therapy with Testoderm testosterone transdermal systems: evaluation of prostatic effects. In *2nd International Androgen Workshop*, Long Beach California, p. 8

20. Krauss, D. J., Taub, H. A. and Lantiga, L. J. (1991). Risks of blood volume changes in hypogonadal men treated with testosterone enanthate for erectile impotence. *J. Urol.*, **146**, 1566–70

DISCUSSION

Schröder: Is it really fair with these small numbers of men studied to conclude that liver toxicity is very unlikely to occur?

Tenover: You are right, in terms of data in older men we do not have enough. Nevertheless, we have data on younger hypogonadal men, and we know that liver function does not change much with age *per se*.

Vermeulen: In Figure 1, testosterone levels in a population we studied are presented. Figure 2 provides the incidences of low testosterone levels in this group. Low testosterone was defined as less than 11 nmols/l which corresponds to about 320 ng/dl. In the group aged 20–40, there was only one 'hypogonadic' patient. However, between 40 and 60 years of age, 60 and 80, and over 80, we found that 7%, 21% and 35% respectively of the men investigated were actually hypogonadic. It must be noted that the study sample comprised 300 presumed healthy men without disease or complaints. I think these data are important in showing us that the frequency of below normal androgen levels is surprisingly high.

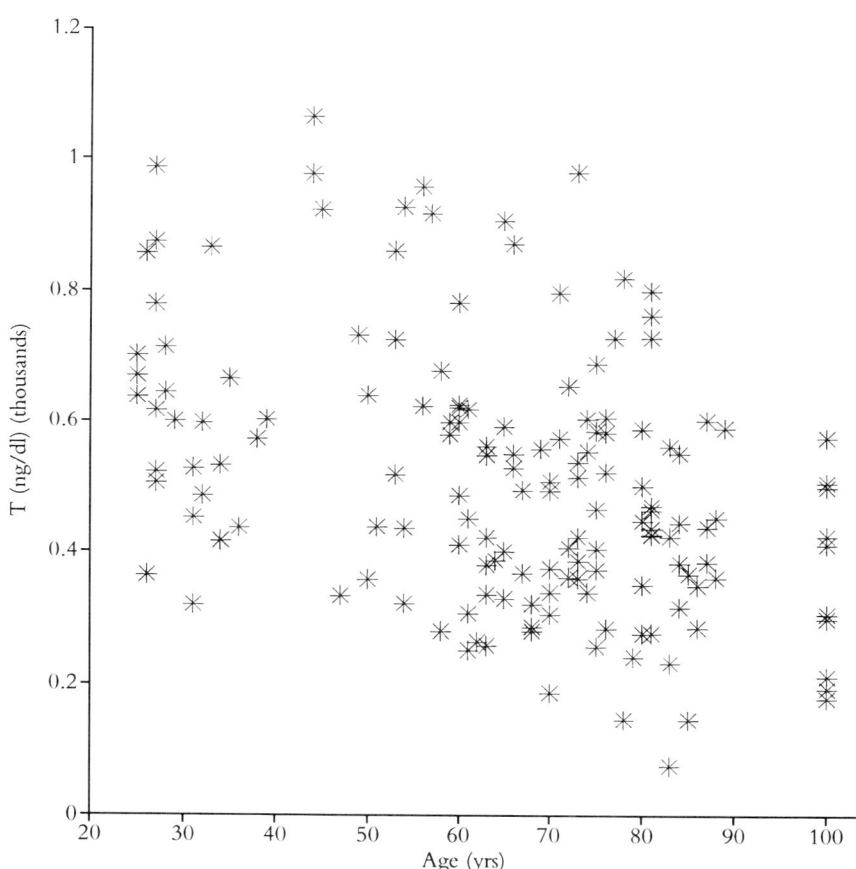

Figure 1   Distribution of testosterone levels in healthy men aged 25–100 in relation to age (*n* = 300). Unpublished findings by Vermeulen and Kaufman. T = testosterone

Morales: Could you advise me on the best timing for determining serum testosterone? The recommendation has been that you should always do it in the morning between 9 and 11. I have run across a number of publications and one study indicated that perhaps only 50% of the population has this circadian variation of high levels in the morning and low levels in the evening, whereas 25% has exactly the reverse pattern, and the rest fairly stable levels throughout the day[1]. Another (Japanese) publication suggested

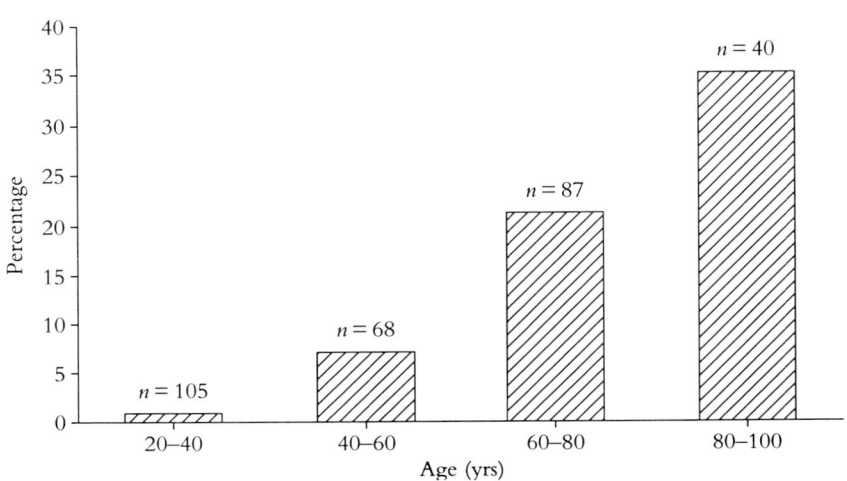

Figure 2 Percentage of testosterone deficient men (< 11 nmol/l) according to age. Unpublished findings by Vermeulen and Kaufman

that the pattern is the reverse in the majority of people[2]. If we define hypogonadic patients exclusively on the basis of biochemical measurements, should we then not question the accuracy of our measurements?

Vermeulen: Two observations might be relevant to your question. It is well documented that the circadian rhythm in aging males is reduced when compared with young men. However, it goes too far to suggest that the rhythm may actually be reversed in some instances. Secondly, I have never seen that there were higher levels in the afternoon than in the early morning. Therefore, I would suggest, at least as far as our Western populations are concerned, that taking blood samples in the morning is indicated.

Gooren: Until what age should we go on with testosterone supplementation in the aging male?

Vermeulen: If the general condition improves with treatment of androgens, I do not see why the treatment should be stopped. Of course, balance must be found between the benefits of the treatment and, for example, difficulties experienced by some patients who have to come each time for follow-up. If they feel better with treatment – if their muscle mass improves and their general well-being or physical activity – I do not see why treatment should be stopped.

Oddens: Dr Tenover, you referred to a few larger androgen supplementation studies which are currently being carried out by various groups including your own. Could you provide some more details of what is being studied, and of what we might expect in terms of results?

Tenover: There are three studies currently being run, by: Marc Blackmon at the Johns Hopkins University in Baltimore; Peter Snyder at the University of Pennsylvania; and myself. Peter Snyder's major outcome parameter is bone, where about 100 men are being studied for a duration of 3 years. We are looking at bone density, biochemical parameters of bone turnover, muscle mass, fat mass, strength and balance. The study duration is also 3 years and we included 70 men. The group at Hopkins is largely interested in cardiac function and muscle, and are investigating protein synthesis, actin and myosin mRNA, etc.; they are also evaluating cognitive function, mood and sleep. Their study design is shorter, running for about 1 year, since they are not primarily looking at outcomes in bone. In the latter study, a group that is receiving testosterone plus growth hormone is compared with either one alone. I believe Marc Blackmon is planning to have a total of around 25 men in each of the groups on testosterone. In terms of results, all we have available now is the safety data, because the studies are blinded and still in progress. From our study, we have observed some changes in hematocrit, which could however be dealt with by decreasing the dose of androgen given. Changes in prostate-specific antigen, urine flow rate and other detection parameters of prostate problems have so far been minimal. Furthermore, cholesterol levels do not seem to increase. In these various respects, the findings accord well with what we would have guessed from the shorter duration studies available to date.

REFERENCES

1. Juenja' H. S., Karanth, S., Dutt' A., Parte' P. and Meherjee, P. (1991). Diurnal variations and temporal coupling of bioactive and immunoreactive luteinizing

hormone, prolactin, testosterone and 17β-estradiol in adult men. *Horm. Res.* **35**, 89–94

2. Yamaguchi, M., Mizunuma, H., Miyamoto, K., Hasegawa, Y., Ibiki, Y. and Igarashi, M. (1991). Immunoreactive inhibin concentrations in adult men: Presence of a circadian rhythm. *J. Clin. Endocr. Metab.*, **72**, 554–9

12

Risks associated with long-term androgen supplementation

L. J. G. Gooren and E. J. Giltay

This chapter deals with the risks associated with long-term androgen supplementation, in particular cardiovascular risks. It focuses on the aging male eligible for androgen supplementation in view of his declining androgen levels in senescence. This chapter will not address the effects of unduly high dosages of androgenic hormones such as those used by bodybuilders.

Unlike thyroid hormones, corticosteroids, or maybe even estrogens, too low or too high serum levels of androgens cause, clinically speaking, few alarming symptoms. With regard to androgens there is no true clinical counterpart of hypo/hyperthyroidism or hypo/hypercortisolism. That might explain the paucity of literature data on side-effects of androgens.

Unlike thyroid hormone and corticosteroid hormone substitution, androgen supplementation modalities are still not satisfactory. Though transdermal patches[1] and even more so subcutaneous testosterone implants[2], constitute veritable improvements, the pharmacokinetics and pharmacodynamics of the more traditional forms of androgen replacement fail to mimic the serum androgen profiles produced by the testis. This situation is reflected in the reported side-effects of androgen administration. Such side-effects might be the result of the modes of testosterone replacement leading to unphysiologically high/low levels of testosterone, 5α-dihydrotestosterone (DHT) or the aromatization product estrogen, and their ratios. Furthermore, the route of administration matters; transdermal administration of testosterone is associated with high DHT levels. This is also the case with the oral testosterone undecanoate. This oral androgen circumvents a first pass through the liver through its absorption from the gut, along with fats, *via* the thoracic duct. Probably a fair amount of testosterone still arrives *via* the portal vein in the liver and causes a significant decline in the production of sex hormone-binding globulin (SHBG)[3]. While it is documented much more clearly for estrogens, it seems likely that androgens also exert different metabolic effects depending on their route of administration, oral or parenteral.

A substantial part of the side-effects observed during androgen administration comes from reports on steroid-abusing sportsmen. Their misuse of androgens does not provide a realistic picture of the side-effects of normal use of androgens.

From the above it follows that at present it is almost impossible to provide an accurate account of side-effects of androgen administration to aging men, in whom only moderate androgen supplementation is needed. The following description of effects of androgens must therefore be viewed in the light of potentiality of such effects of treatment. As stated above, much of the information about the effects of androgens on biological systems has been learned from an unphysiological context and it remains to be established whether these side-effects occur at all or are biologically significant with the moderate supplementation doses suited in senescence.

In contrast to replacement with thyroid hormone or corticosteroids, suitable valid clinical and biochemical indices of optimal testosterone supplementation have not been established in the scientific literature. This makes studies that aim to define beneficial and deleterious effects of androgen supplementation still a difficult undertaking.

SIDE-EFFECTS OF ANDROGEN ADMINISTRATION

Our clinic provides comprehensive care to transsexual patients; this includes cross-sex hormonal treatment. Between 1975 and 1995, 293 female-to-male transsexuals have received treatment with androgens. Their ages ranged from 17 to 70 years with a mean age of 34 years. The total duration of exposure to androgens of this group amounted to 2418 years. Subjects were treated most of the times with parenteral androgens (Sustanon-250® (Organon) or Testoviron-180® (Schering Pharma)/2 weeks) and sometimes with oral androgens (testosterone undecanoate (Andriol®, Organon) four to six capsules of 40 mg per day).

Several of our studies with regard to cardiovascular risks of androgens have been performed in this group of subjects and will be mentioned below. To our relief, we were unable to find a higher degree of mortality and morbidity in this group of androgen-treated subjects in comparison with age-matched subjects. In 12 subjects elevated blood pressure was found, which is in accordance with the expected number on the basis of prevalence studies in the general population. In 45 subjects elevations of liver enzymes were found, but they never exceeded 2.5 times the upper normal limit. In 40 cases no other explanation than androgen administration could be found. In 20 cases this lasted for less than 6 months, in the remaining 20 longer but this did not require discontinuation of androgen administration. Sodium and water retention was observed in five cases but was corrected with a dose reduction of the androgens. Severe acne was observed in 80 cases, predominantly on the upper back as is the case in men starting androgen administration rather late in life. This problem could be acceptably treated with topical agents.

Of 86 transsexual females treated with androgens, data were collected on their weight changes in the 1st year of hormone administration. There was an increase of on average 5.2 kg (range 1–14 kg) in weight in 69 subjects, a decrease of on average 5.6 kg (range 4–12 kg) in eight subjects, and in nine subjects no weight changes were recorded. Part of the weight gain must be explained by an increase in muscle mass, on average 1.2 kg in this group of androgen-naive subjects.

In conclusion: our data on side-effects of androgen administration to this particular group of subjects were reassuring that androgens are acceptably safe drugs.

ANDROGENS AND CARDIOVASCULAR EFFECTS

Androgens and vasoactive substances

Among the most significant vasoactive substances are the endothelins. Endothelins are a family of peptides which are produced in a variety of tissues where they act as modulators of vasomotor tone, cell proliferation and hormone production. Studies with endothelins and specific endothelin-receptor antagonists have suggested that these peptides are important in vascular physiology and disease. Endothelins should be regarded more as paracrine than endocrine hormones. The production of endothelins is regulated by a variety of hormones, other vasoactive substances and conditions of vascular stress. Endothelin-1 probably has a role in the maintenance of basal vasomotor tone and it is a very potent vasoconstrictor, counteracted by nitric oxide, prostacyclin and atrial natriuretic peptide.

It is likely that endothelin-1 plays a role in atherosclerosis and cardiac hypertrophy (for a review see Levin[4]).

We found that testosterone administration to female-to-male transsexuals increased plasma endothelin levels, while androgen deprivation with cyproterone acetate with simultaneous administration of ethinyl estradiol decreased endothelin levels[5]. Indeed, in a study of female cynomolgus monkeys it was found that induction of male plasma androgen levels was strongly atherogenic independently of variations in plasma lipoprotein and non-lipoprotein risk variables, though it reversed atherosclerosis-related impairment of endothelium-dependent vasodilator responses in this study[6]. It is too early to determine whether replacement with physiological doses of androgens does indeed contribute to cardiovascular risks through an endothelin-mediated mechanism, but it seems advisable to pay attention to this factor in future studies of androgen replacement in elderly men.

Other factors to be considered are the prostaglandins. Platelet aggregation is prevented by prostaglandin I_2 (PGI$_2$, prostacyclin), a vasodilatatory and antiaggregatory eicosanoid, and is stimulated by thromboxane B_2 (TXB$_2$), a vasoconstricting and proaggregatory eicosanoid. It has been shown that testosterone inhibits PGI$_2$ production in aortic smooth muscle cell cultures, whereas testosterone significantly increased TXB$_2$ in monkeys[7]. Furthermore, the expression of TXA$_2$ receptors is up-regulated by testosterone, which may contribute to the thrombogenicity of androgenic steroids[8]. There have indeed been incidental reports of thrombotic complications associated with androgen administration[9,10]. Such effects were more apparent when high-dose anabolic steroids were used[11]. By contrast, administration of anabolic 17α-alkylated androgens may increase fibrinolytic activity and the natural anticoagulant proteins, protein C, protein S[12] and antithrombin III[13], thereby partly counterbalancing the unfavorable effects of androgens. These observations suggest strongly that, apart from effects on coagulation, vascular factors play a significant role in the increased thrombosis risks of androgen administration.

Androgens and sodium retention

All androgens cause some degree of sodium retention and expansion of extracellular fluid volume[14]. This effect is usually small; weight increases by about 3% in healthy individuals. However, in patients with heart and/or renal disease an already fragile balance may be disturbed, and edema may develop.

Androgens and body fat distribution

From puberty onwards males and females differ in their body distribution of fat over the subcutaneous and intra-abdominal (or visceral) fat deposits. Subcutaneous fat disappears largely in normal weight boys and visceral fat stores are also small but start to become manifest with increasing age. From age 20–30 years onwards the large visceral fat storage of men is twice as large as that of women with a comparable body mass index. If men store fat, they accumulate it preferentially intra-abdominally, while in women this occurs preferentially subcutaneously on hips and thighs.

That androgens are involved in this sex-specific fat distribution pattern has become increasingly clear though the precise mechanism remains unknown. Women with elevated levels of serum androgens have more visceral fat than control women. In our own experiments we studied the effects of androgen administration on fat distribution in females in female-to-male transsexuals undergoing sex reassignment. It could be shown that androgens lead to a quick depletion of subcutaneous fat, while it takes between 1 and 3 years for visceral fat accumulation to become manifest.

Visceral fat accumulation has been found to be a risk factor for cardiovascular disease and for non-insulin-dependent diabetes mellitus. Its mechanism has not been fully clarified but it has been found that visceral fat is metabolically very active with a high turnover of triglycerides. The enlarged abdominal adipocytes show an enhanced sensitivity to lipolytic stimuli resulting in an increased production of free fatty acids. It has been hypothesized that testosterone stimulates lipolysis in the visceral fat deposits. The subsequent load of triglycerides and/or free fatty acids reaching the liver *via* the portal vein interferes with the hepatic handling of insulin leading to hyperinsulinism and a deteriorated glucose tolerance, and further, to a lowering of high-density lipoprotein (HDL) cholesterol and higher serum levels of triglycerides. This constitutes a risk for cardiovascular disease.

Though the above-described visceral fat accumulation seems to be androgen-related, studies have found that men with visceral accumulation of body fat often have relatively low testosterone; the mechanism of this non-linear association has to be elucidated.

The first experiments testing whether androgen supplementation in middle-aged obese men has a beneficial effect on visceral fat deposits, insulin sensitivity and serum lipids have been carried out. Marin and colleagues[15] found such an effect in a group of 23 men who were treated with oral testosterone undecanoate 80 mg/day for 8 months inducing a decline in levels of SHBG and therewith probably a rise of free testosterone levels. In a later study at the same laboratory, it was found that the turnover rate of depot triglyceride was more rapid in abdominal compared to femoral adipose tissue. Testosterone administration inhibited triglyceride uptake and lipoprotein lipase activity and caused a more rapid turnover of triglycerides only in the abdominal adipose tissue region. These results demonstrated the marked effects of testosterone on adipose tissue metabolism and suggested that testosterone is an important regulator of the proportion of depot fat mass in central and peripheral adipose tissue in men[16]. In the study of Tenover[17] where testosterone enanthate (100 mg/week) was given for 3 months, total body weight and lean body mass increased, but body fat and waist to hip ratio (an index of the relation of the amount of abdominal and subcutaneous fat) did not change.

Androgens and serum lipids

There is no consensus on the effect of androgens on serum lipids and lipoproteins. This is (partially) due to differences in study designs and whether endogenous or exogenous androgen effects are investigated. The picture is even more complicated to evaluate since part of both endogenous and exogenous testosterone is metabolized to estrogens.

There is, however, consensus that the incidence of coronary disease is significantly higher in men than in women until menopause after which the incidence becomes similar. Therefore there is at first sight reason to believe that this sex difference in incidence of coronary heart disease is (in part) related to differences in the endocrine milieu causing differences in lipid profiles. Serum HDL cholesterol is lower and triglyceride levels are higher in eugonadal men than in premenopausal women while these differences are not manifest before puberty.

Very convincing evidence that testosterone is involved in this sex difference comes from experiments wherein androgen levels were severely lowered whereupon HDL cholesterol levels increased. When serum androgen levels were subsequently returned to normal, serum levels of HDL cholesterol decreased to normal male levels. Most, but not all, studies in adult men find a positive correlation between plasma testosterone and plasma HDL cholesterol concentrations. Administration of aromatizable androgens to hypogonadal men caused an increase in HDL cholesterol, while non-aromatizable androgens did the opposite. In our own study the oral (aromatizable) androgen testosterone undecanoate was administered to hypogonadal men and to previously non-treated female-to-male trans-sexuals. While serum estradiol levels in the females were three to four times higher during testosterone administration, in both sexes levels of HDL cholesterol and HDL$_2$ cholesterol declined and were eventually of the same magnitude. This led us to conclude that testosterone is indeed the major determinant of the sex difference in HDL cholesterol levels[18]. In line with this finding is that women with hyperandrogenism show 'male-like' plasma lipoprotein profiles[6,19].

Some studies in aging men have shown results that seem to contradict the overall notion that androgens by their action on lipid profiles would increase the risk for coronary artery disease. In a study of men with non–insulin-dependent diabetes mellitus these patients showed lower levels of endogenous total and free testosterone; their HDL cholesterol levels were lower but triglyceride levels higher than in controls[20]. In support of this finding another study found that in men with coronary artery disease plasma levels of testosterone correlated positively with HDL cholesterol and negatively with the cardiovascular risk factors fibrinogen, plasminogen activator inhibitor-1 and insulin levels[21]. In a study of geriatric male patients who had suffered a myocardial infarction it was found that these patients had low testosterone levels in a threshold manner[22]. These studies suggest the intriguing possibility that, in spite of the overall negative effects of androgens on lipid profiles, a lower-than-normal androgen level in aging men is associated with an increase of atherosclerotic disease.

The first results of studies wherein testosterone was actually administered to aging men with low testosterone levels were comforting. In a double-blind, placebo-controlled, cross-over study Tenover[17] found that adminis-tering testosterone enanthate (100 mg/week) for 3 months to 13 healthy elderly men with low serum total and non-sex hormone-binding globulin-bound testosterone levels, decreased total and low density lipoprotein (LDL) cholesterol without changing HDL cholesterol. In agreement with these results are the findings of Morley and co-workers[23]. Administration of 200 mg testosterone enanthate every 2 weeks for 3 months decreased total cholesterol without changing HDL cholesterol levels. Of great interest is the study of Ellyin[24] which established, in a 2-year study of administration of testosterone cypionate (25 mg/week or 50 mg/2 weeks) to elderly men with low testosterone levels, a decrease in total and LDL cholesterol and no change in HDL cholesterol and triglycerides.

In summary, not all results of studies on the relationship between testosterone and lipid profiles indicate that androgens induce a more atherogenic lipid profile; actually, the studies of testosterone administration in aging men point in another direction. When studying the effects of androgens on lipid profiles in relation to cardiovascular disease, it has to be kept in mind that androgens also have direct effects on the vascular wall and on insulin sensitivity independent of their effects on lipids and the latter could also be invoked to explain sex differences in coronary artery disease.

Androgens and insulin resistance

Evidence that insulin resistance induces an unfavorable lipid profile, a decrease in plasma HDL cholesterol and an increase in plasma triglycerides, has become quite convincing[25]. There is very likely a link between hyper-insulinemia (or insulin resistance, with which it is inextricably associated) and hypertension in otherwise healthy persons in whom high plasma insulin levels are the only abnormality[26,27]. Coexistence of insulin resistance and hyperandrogenism has been described frequently, mainly in women with polycystic ovary syndrome. An improvement of insulin resistance could be observed when the hyperandrogenism was corrected in these women[28]. Several reports describe induction of insulin resistance in men taking anabolic steroids[29]. However, others have suggested that androgen treatment in elderly males actually improves insulin resistance.

In our own experiments, using the hyperinsulinemic–euglycemic clamp method, we studied the effects of the administration of 250 mg testosterone esters/2 weeks on insulin resistance in a group of female-to-male transsexuals[30]. In hyperinsulinemic–euglycemic clamp studies insulin is infused at pre-set standardized infusion rates, and the amount of glucose required to keep blood glucose levels constant provides a reliable indication of insulin sensitivity or, alternatively, of insulin resistance, of a particular subject. In our study glucose utilization decreased significantly with testosterone administration showing that less glucose entered the cells; in other words, insulin became biologically less effective in its ability to deliver glucose to the cell, i.e. the development of peripheral insulin resistance. The effects of androgens were clearest when insulin levels in the clamp study were in the physiological range. Glucose utilization rates were much less affected with higher (supraphysiological) insulin levels; higher than normal levels of insulin could apparently compensate the resistance observed with normal insulin levels. This must be interpreted as indicating that insulin responsiveness at the postreceptor level was not seriously diminished. Insulin could still exert its biological action, be it at higher hormone concentration. Further, the insulin-mediated inhibition of endogenous glucose production by the liver was not affected by androgen administration, showing that the effects on insulin sensitivity occurred only at insulin's peripheral site of action.

Androgens and hematopoiesis and sleep apnea

From puberty on men have higher hemoglobin levels, hematocrits and red blood cell counts than women. Both testosterone and 5α-dihydrotestosterone stimulate renal production of erythropoietin. There is evidence for a direct effect of androgens on erythropoietic stem cells[31]. Androgen receptors have been found in cultured erythroblasts[32]. Young hypogonadal men have lower red blood cell counts and hematocrits than age-matched controls. These values increase upon administration of androgens to hypogonadal men[33]. Healthy older men tend to have similar or slightly lower hematocrit values than young adult men[31]. Two studies of elderly men receiving androgen administration have shown increases of hematocrit of up to 7%[17,23].

In a study of Matsumoto and co-workers[33] it was shown that androgen levels may play an important role in the pathogenesis of obstructive sleep apnea and that this may be a complication of testosterone therapy, though the relevance of androgens for obstructive sleep apnea could not be shown in a study wherein the pure antiandrogen flutamide was used[34]. Since these apneic events and the ensuing oxygen desaturation may lead to cardiovascular complications, it is pertinent to ask older men about this symptom and to measure and follow up hematocrit values before and during androgen administration. Care must be exercised in men who are overweight, heavy smokers, or who have chronic obstructive airway disease.

In humans androgens appear to enhance platelet aggregation, through a mechanism involving substances secreted by the vascular wall, which was dealt with in a previous section.

ANDROGENS AND THE LIVER

Most clinicians will encounter few side-effects of present androgen treatment on the liver. Several of the more serious side-effects of androgens on the liver (such as peliosis hepatis, subcellular changes of hepatocytes, hepatocellular hyperplasia and hepatocellular adenomas) turned out to be largely, though not exclusively, associated with the use of 17α-alkylated androgens or abuse of anabolic steroids. There is presumably a relation between side-effects of androgens on the liver and previous liver integrity. With a modest androgen supplementation to the aging male, no serious side-effects need to be expected.

ANDROGENS AND GYNECOMASTIA

Gynecomastia develops in cases of an increased estrogen/androgen ratio, or more precisely an estrogen/androgen ratio that effectively acts at the breast tissue (for a review see Glass[35]). The aging process itself is associated with a loss of androgen production, a rise of SHBG and an increase of adipose tissue, the site of aromatase activity enabling the conversion of androgens to estrogens. The combination of these factors leads already to a prevalence of spontaneous gynecomastia in old age. Administration of aromatizable androgens may lead to gynecomastia in both puberty and old age, in all likelihood when the subsequent increases of estrogens and androgens tip the balance in favor of a higher estrogen/androgen ratio. This side-effect is medically innocent and may subside over time. Patients with liver or renal disease may be particularly predisposed to the development of gynecomastia.

ANDROGENS AND SKIN EFFECTS

Androgens have a profound effect on the skin. Androgens regulate, by means of receptors, the sebaceous glands and hair growth function, and remarkably they do so on two different types of target cells in the skin: the epithelial sebocyte of the sebaceous gland and the mesenchymal cells of the hair follicle dermal papilla. Excessive oiliness of the skin and the development of acne are seen in younger men given androgen replacement. Whether these skin changes will occur in elderly men is not known. Middle-aged men starting on androgen replacement almost never show acne. In genetically predisposed individuals androgens induce male pattern alopecia.

ANDROGENS AND INTERACTION WITH OTHER DRUGS

Few interactions of androgens with other drugs have been reported. In patients using anticoagulants, anabolic steroids may reduce the dose which a patient requires by up to 25%[36]. This in part is due to the effect on coagulation factors; they may decrease the synthesis or increase the degradation of coagulation factors. Also, an increase of the natural anticoagulant antithrombin III has been found. Although not all patients are affected equally, this should be taken into account when anticoagulants are given to patients using androgenic hormones, requiring a dose reduction of anticoagulants in these patients. There is some evidence from *in vitro* and animal experiments that the inhibitory effect of aspirin on platelet aggregation occurs only in the presence of androgens, an effect distinct from the above action of androgens on coagulation factors[37].

ANDROGENS AND PSYCHOLOGICAL FUNCTIONS

Reliable studies on the relationships between androgens and psychological functions are of rather recent date. There is now solid evidence that androgens stimulate sexual desire. With regard to erectile function the situation is somewhat less clear. Nocturnal and spontaneous erections are clearly androgen dependent; erections in response to visual erotic stimuli are less androgen dependent though achievement of erections and rigidity show a relationship; remarkably, detumescense appears to be prolonged with lower testosterone levels. Erections in response to auditory stimuli and fantasy are probably androgen dependent. It has become clear that in males between 20 and 50 years of age, approximately 60–80% of the normal physiological levels suffice to maintain sexual functions and that increasing testosterone levels above that threshold adds little to sexual functioning. Whether this holds true for aging men, remains to be established. Some believe that the sensitivity to androgens decreases with aging[38]. But it seems safe to assume that a modest increase of testosterone levels in aging men will not lead to sexually deviant behavior. The relationship between testosterone and aggression is less well researched, but a relationship with anger proneness and tendencies to aggression is demonstrable. The latter is not identical with overt aggression. Recently there has been a large number of reports on the psychological effects of anabolic steroids. These may vary from depression to mania to overt violence specifically in response to provocation. It is likely that these effects are related to the doses used and are unlikely to occur with moderate androgen supplementation.

CONCLUSION

The first reports on androgen supplementation to aging men are rather positive[17,23,24]. It is at present difficult to predict whether androgen supplementation with modest dosages carries a significant risk. A reliable evaluation of that risk is hampered by our less than ideal androgen treatment modalities. Ideally, they should provide the aging male with only that amount of testosterone that restores his androgen deficiency.

With regard to cardiovascular risk factors there are literature data to suggest that androgens influence a number of risk factors in a negative direction, while others suggest the reverse. There is no consensus on the negative effects of androgens on lipid profiles and on body fat distribution. Androgens seem to induce insulin resistance but may also improve it, and may alter vascular factors to a higher risk profile. There is at this moment no clear evaluation possible as to how much these observed shifts in cardiovascular risk factors will in fact contribute to actual cardiovascular morbidity and mortality. In view of the positive effects of the above mentioned studies, further investigations are warranted. Analogous to the studies on estrogen replacement in postmenopausal women, well-designed, long-term studies to monitor beneficial and negative effects are needed.

ACKNOWLEDGEMENT

Dr C. D. A. Stehouwer is thanked for his valuable comments on the manuscript.

REFERENCES

1. Meikle, A. W., Mazer, N. A., Moellmer, J. F., Stringham, J. D., Tolman, K. G., Sanders, S. W. and Odell, W. D. (1992). Enhanced transdermal

delivery of testosterone across nonscrotal skin produces physiological concentrations of testosterone and its metabolites in hypogonadal men. *J. Clin. Endocrinol. Metab.*, **74**, 623–8

2. Handelsman, D. J., Conway, A. J. and Boylan, L. M. (1990). Pharmacokinetics and pharmacodynamics of testosterone pellets in man. *J. Clin. Endocrinol. Metab.*, **71**, 216–22

3. Conway, A. J., Boylan, L. M., Howe, C., Ross, G. and Handelsman, D. J. (1988). Randomized clinical trial of testosterone replacement therapy in hypogonadal men. *Int. J. Androl.*, **11**, 247–64

4. Levin, E. R. (1995). Endothelins. *N. Engl. J. Med.*, **333**, 356–63

5. Polderman, K. H., Stehouwer, C. D. A., Van de Kamp, G. J., Dekker, G. A., Verheugt, F. W. A. and Gooren, L. J. G. (1993). Influence of sex hormones on plasma endothelin levels. *Ann. Int. Med.*, **118**, 429–32

6. Adams, M. R., Williams, J. K. and Kaplan, J. R. (1995). Effects of androgens on coronary artery atherosclerosis and atherosclerosis-related impairment of vascular responsiveness. *Arterioscler. Thromb. Vasc. Biol.*, **15**, 562–70

7. Derman, R. J. (1995). Effects of sex steroids on women's health: implications for practitioners. *Am. J. Med.*, **98** (Suppl. 1A), 137S–43S

8. Ajayi, A. A. (1995). Testosterone increases human platelet thromboxane A2 receptor density and aggregation responses. *Circulation*, **91**, 2742–7

9. Nagelberg, S. B., Laue, L., Loriaux, D. L., Liu, L. and Sherin, R. J. (1986). Cerebrovascular accident associated with testosterone therapy in a 20-year old hypogonadal man (letter). *N. Engl. J. Med.*, **314**, 649–50

10. Shizowa, Z., Yamada, H., Mabuchi, C., Hotta, T., Saito, M., Sobue, I. and Huang, Y. P. (1982). Superior sagittal sinus thrombosis associated with androgen therapy. *Ann. Neurol.*, **12**, 57–88

11. Mochizuki, R. M. and Richter, K. J. (1988). Cardiomyopathy and cerebrovascular accident associated with anabolic androgenic steroid use. *Phys. Sportsmed.*, **16**, 109–14

12. Ansell, J. E., Tiarks, C. and Fairchild, V. K. (1993). Coagulation abnormalities associated with the use of anabolic steroids. *Am. Heart J.*, **125**, 367–71

13. Ammus, S. (1989). The role of androgens in the treatment of hematologic disorders. *Adv. Intern. Med.*, **34**, 191–208

14. Wilson, J. D. (1987). Androgen abuse by athletes. *Endocr. Rev.*, **2**, 181–99

15. Marin, P., Holmäng, S. Jönsson, L., Sjöström, L., Kvist, H., Holm, G., Lindstedt, G. and Björntorp, P. (1992). The effects of testosterone on body composition and metabolism in middle-aged obese men. *Int. J. Obesity*, **16**, 991–7

16. Marin, P., Odén, B. and Björntorp, P. (1995). Assimilation and mobilization of triglycerides in subcutaneous abdominal and femoral adipose tissue *in vivo* in men: effects of androgens. *J. Clin. Endocrinol. Metab.*, **80**, 239–43

17. Tenover, J. S. (1992). Effects of testosterone supplementation in the aging male. *J. Clin. Endocrinol. Metab.*, **75**, 1092–8

18. Asscheman, H., Gooren, L. J. G., Megens, J. A. J., Nauta, J., Kloosterboer, H. J. and Eikelboom, F. (1994). Serum testosterone level is the major determinant of the male-female differences in serum levels of high density lipoprotein cholesterol and HDL2 cholesterol. *Metabolism*, **43**, 935–9

19. Plymate, S. R. and Swerdloff, R. S. (1992). Androgens, lipids and cardiovascular risk. *Ann. Intern. Med.*, **117**, 871–2

20. Barrett-Connor, E. (1992). Lower endogenous androgen levels and dyslipidemia in men with non-insulin-dependent diabetes mellitus. *Ann. Intern. Med.*, **117**, 807–11

21. Phillips, G. B., Pinkernell, B. H. and Jing, T-Y. (1994). The association of hypotestosteronemia with coronary artery disease in men. *Arterioscler. Thromb.*, **14**, 701–6

22. Swartz, C. A. and Young, M. A. (1987). Low serum testosterone and myocardial infarction in geriatric male inpatients. *J. Am. Geriatr. Soc.*, **35**, 39–44

23. Morley, J. E., Perry, III. H. M., Kaiser, F. E., Kraenzle, D., Jensen, J., Houston, K., Mattammal, M. and Perry Jr. H. M. (1993). Effects of testosterone replacement therapy in old hypogonadal males: a preliminary study. *J. Am. Geriatr. Soc.*, **41**, 149–52

24. Ellyin, F. M. (1995). The long term beneficial effect of low dose testosterone in the aging male. In *The Endocrine Society 77th Annual Meeting*. Program and Abstracts, p. 322, P2–127 (Washington DC: The Endocrine Society Press)

25. Reaven, G. M. (1988). Banting lecture 1988: role of insulin resistance in human disease. *Diabetes*, **37**, 1595–607

26. Zavaroni, I., Dall'Aglio, E., Alpi, O., Bruschi, F., Bonora, E., Pezzarossa, A. and Butturini, U. (1985). Evidence for an independent relationship between plasma insulin and concentrations of high density lipoprotein cholesterol and triglyceride. *Atherosclerosis*, **55**, 259–66

27. Laws, A. and Reaven, G. M. (1993). Insulin resistance and risk factors for coronary heart disease. *Clin. Endocrinol. Metab.*, **7**, 1063–78

28. Shoupe, D. and Lobo, R. A. (1984). The influence of androgens on insulin resistance. *Fertil. Steril.*, **41**, 385–8

29. Cohen, J. C. and Hickman, R. (1987). Insulin resistance and diminished glucose tolerance in powerlifters ingesting anabolic steroids. *J. Clin. Endocrinol. Metab.*, **64**, 960–3

30. Polderman, K. H., Gooren, L. J. G., Asscheman, H., Bakker, A. and Heine, R. J. (1994). Induction of insulin resistance by androgens and estrogens. *J. Clin. Endocrinol. Metab.*, **79**, 265–71

31. Shahidi, N. T. (1973). Androgens and erythropoiesis. *N. Engl. J. Med.*, **289**, 72

32. Claustres, M. and Sultan, C. (1988). Androgen and erythropoiesis: evidence for an androgen receptor in erythroblasts from human bone marrow cultures. *Horm. Res.*, **29**, 17

33. Matsumoto, A. M., Sandblom, R. E., Schoene, R. E., Lee, K. A., Giblin, E. C., Pierson, D. J. and Bremner, W. J. (1985). Testosterone replacement in hypogonadal men. *Clin. Endocrinol.*, **22**, 713–21

34. Stewart, D. A., Grunstein, R. R., Berthon-Jones, M., Handelsman, D. J. and Sullivan, C. E. (1992). Androgen blockade does not affect sleep-disordered breathing or chemosensitivity in men with obstructive sleep apnea. *Am. Rev. Resp. Dis.*, **146**, 1389–93

35. Glass, A. R. (1994). Gynecomastia. *Endocrinol. Metab. Clin. North Am.*, **23**, 825–37

36. Westerholm, B. (1988). Sex hormones. In Dukes, M. N. G. (ed.) *Meyler's Side Effects of Drugs*, Chapter 42b, pp. 866–7 (Amsterdam: Elsevier)

37. Spranger, M., Aspey, B. S. and Harrison, M. J. (1989). Sex difference in antithrombotic effects of aspirin. *Stroke*, **20**, 34–7

38. Schiavi, R. C. (1990). Sexuality and aging in men. *Annu. Rev. Sex. Res.*, **1**, 227–50

DISCUSSION

Vermeulen: You observed relatively often elevations of liver enzymes in the female to male transsexuals treated. Was there a difference in this respect between subjects treated with oral testosterone *versus* those treated with intramuscular or parenteral testosterone?

Gooren: When we evaluated liver enzymes, all the subjects were using parenteral testosterone. Our policy is to start all of them on parenteral testosterone, with the aim of suppressing their menstrual activity. Suppression of menstrual activity does not happen if you administer oral testosterone, unless you add a progestogen. There were a number of subjects in whom we could explain the liver function disturbances observed: hepatitis, alcoholism, etc., but in a large number, we had no other explanation than androgen administration. However, in this respect we could not compare oral administration with parenteral, since all used parenteral testosterone.

Orwoll: I was intrigued by the fact that subcutaneous fat changed in your female to male transsexuals, and that this was a transient change, which later disappeared. I wonder if you have a comment about that?

Gooren: Let me say first that they virtually all gained weight. The fat had to go somewhere, so to speak, and it went preferentially in the abdomen once they were using androgens, but if there was no longer room for storage, it went subcutaneously again.

Orwoll: As a corollary, was the visceral fat in male to female transsexuals altered with estrogen therapy? In other words, did visceral fat go away in these subjects?

Gooren: It seems very much like that, but our research in this connection was less sophisticated than that in the female to male transsexuals. This type of research is rather expensive. You have to find funding for it and I must say that, as soon as you mention the word 'transsexual' in a grant application, there is a fair chance that it does not get funded.

Schröder: What happened to the libido in the male to female transsexuals who received estrogens? Does the normal male libido disappear, is it replaced by a female libido – do they continue being sexually active?

Gooren: Transsexuals might not be the right subjects to study this question, but what you clearly see is a decrease in sexual interest in male to females on estrogen treatment. They lose much of their interest and claim that their sexuality become more female, leaving them less interested in performance and more in relationship aspects. In female to male transsexuals you see a clear increase in libido and sexual interest, and an interest in one would almost say 'quick sex'. So these effects are clearly there in the transsexual population. But sometimes you get very surprised: in our population of female to male transsexuals, there are two people with androgen insensitivity. These two subjects with almost no androgen activity are the subjects with probably the highest libido in the total study population. Such perhaps exceptional observations make me always question the relevance of androgens for sexual interest, libido, etc.

Tenover: I sometimes find that clotting parameters, vasodilator and vasoconstrictor levels, etc. are a very confusing issue. However, it is actually very relevant. Would you say that the vasoconstrictive effects and clotting effects are sort of balanced out by other effects during androgen administration?

Gooren: I think it is indeed kind of balanced out. What you see is that the Von Willebrand factor decreases with androgens, which is beneficial whereas the endothelins increase, and that is negative. It is difficult to make a strong statement about the other activators and inhibitors. However, the changes may be overall neutral.

13

The prostate and androgens: the risks of supplementation

F. H. Schröder

PROSTATE CANCER GROWTH IS STIMULATED BY ANDROGENS

Manifest, metastatic or locally confined prostate cancer is usually dependent on androgens in its growth. Withdrawal of androgens leads to objective response rates of metastatic lesions and of the primary tumor which are in the range of 40–80%. Subjective response rates may even be higher. Response to antiandrogenic endocrine treatment can be monitored by tumor markers such as prostate-specific antigen and acid phosphatase, one of the oldest marker substances available in oncology. These two markers, and also alkaline phosphatase if it is elevated, will show a decrease in their serum levels under antiandrogenic treatment.

The process of growth and marker inhibition by androgen withdrawal can be turned around; prostate cancer can be stimulated to grow by endogenous and exogenous androgens. This will be accompanied by an almost simultaneous rise in marker substances. Although the evidence for this phenomenon is relatively scarce, some examples can be given. It has been attempted to use clinical stimulation of prostate cancer to synchronize prostate cancer cell populations in order to increase the effect of cytotoxic therapy by Manni and colleagues[1]. In a similar way, Fowler and Whitmore[2] have attempted to utilize androgenic stimulation of metastatic prostate cancer to increase the amount of [96]strontium incorporation into the prostate cancer cells for therapeutic purposes. In both instances evidence of stimulation was produced. Experience with intermittent antiandrogenic endocrine therapy with cyproterone acetate reported by Goldenberg and co-workers[3] shows that endogenous testosterone also stimulates manifest prostate cancer to grow. Obviously, the quantitative relationship between amounts of androgens and the level of stimulation is difficult to establish in humans. In a few instances human prostate cancer has been transplanted successfully into nude mice. Androgen-dependent lines resulted, in which androgen stimulation and withdrawal could be studied in a dose-dependent fashion by van Weerden and colleagues[4–6]. Respective experiments allow correlation of plasma and intratumoral androgen levels. The results show clearly that there is a strict dose-dependence of androgenic stimulation. Cessation of tumor growth results from castrate levels, total withdrawal of androgens was associated with a more pronounced decrease of tumor volume. Similar experiments were carried out with two different hormone-dependent lines, PC-82 and PC-EW. Tumor growth rates are strictly correlated to plasma and tumor levels of testosterone and 5α-dihydrotestosterone.

In dealing with the subject of this contribution it is crucial to define 'clinically manifest prostate cancer'. Contrary to carcinoma of the breast, many precursor lesions of prostate cancer are known which are very prevalent from age 35–40, as was recently shown by Sakr and colleagues[7].

INITIATION AND PROMOTION

Based on epidemiological and autopsy studies in the pathogenesis of human prostate cancer, a phase of initiation and promotion can be differentiated. While the morphological steps in the development of prostate cancer are well described, these have not yet been correlated sufficiently with genetic alterations to allow a complete picture to be drawn of the biological progression of this disease. Several autopsy studies have revealed a strong age-dependence of the prevalence of human prostate cancer and respective precursor lesion which averages 33% for the age groups of 60–70 years in the series of Franks[8]. Mainly by means of the volume of these lesions and their histological appearance, precursor lesions (prostatic intraepithelial neoplasia, PIN) and atypical adenomatous hyperplasia (AAH) can be differentiated. The relationship of these lesions with the exceedingly frequent focal carcinoma is not clearly established. Focal carcinoma is usually well differentiated and characterized by a very small volume (average 0.02 ml). The majority of lesions detected at autopsy fall in this category. There is no geographic variation in the autopsy prevalence of these focal tumors throughout the world, as was elegantly shown in an autopsy study carried out by Breslow and colleagues[9] in seven different areas of the world. Clinically unsuspected, usually focal, prostate cancer is found at the time of transurethral resection or open prostatectomy for benign prostatic hyperplasia in 8–12% of cases. With advanced screening technology the percentage of such small lesions identified within screen-detected cancers probably amounts to 8–10%. On the whole, with the same technologies, which include prostate-specific antigen, rectal examination and transrectal ultrasonography screening, a detection rate of 4–5% has been described in men between 55 and 75 years[10]. Breslow and colleagues[9], in confirming the results of Akazaki and Stemmermann[11], have shown that precursor lesions are equally frequent in geographic areas with a high and low incidence of clinical prostate cancer. More advanced local lesions found in autopsy series are more frequent in areas of the world with a high incidence of clinical prostate cancer. Areas with a high incidence are characterized by what is called a 'Western lifestyle' and Western dietary habits as opposed to the more vegetarian and low fat diet in those Eastern countries and Mediterranean countries with a low incidence of clinical prostate cancer, as described by Adlercreutz[12].

Morphologically, the phase of initiation is characterized by the origination of precursor lesions and the exceedingly frequent focal carcinoma. It is unclear at this moment which factors lead to promotion of these lesions to larger, more aggressive and potentially deadly prostate cancer in Western countries. In spite of a very strong epidemiological lead it is still unclear which factors govern the promotion of prostate cancer. Candidates are diet, inheritance and endocrine factors.

Several research groups have produced evidence that dietary habits may be correlated with significant changes in plasma levels of male and female steroid hormones. In a prospective carefully planned study in healthy volunteer men, Hämäläinen and co-workers[13] showed that the temporary use of a semi-vegetarian diet will lead to a significant decrease of plasma testosterone levels. De Jong and colleagues[14], as the result of the Dutch–Japanese case–control study of prostate cancer, have shown that age-corrected levels of plasma testosterone and estradiol are lower in elderly men in Japan than in the Netherlands. Unfortunately, these differences were not revealed in comparing cases and controls in this study, possibly because of low statistical power. Higher plasma androgen levels were also not found for prostate cancer cases in uncontrolled comparative studies. One example is the report by Carter and co-workers[15] who did not find elevated plasma levels of luteinizing hormone, testosterone, dihydrotestosterone and SHBG with and without age adjustment for 20 prostate cancer cases compared with age-matched controls, despite the fact that they were able to study plasma levels 0–5, 5–10 and 10–15 years prior to diagnosis. Still, the findings described offer a possibility of linking dietary and endocrine factors in considering the issue of promotion of focal lesions to clinical prostate cancer. If this correlation could be confirmed, this would indicate that small differences in circulating androgen levels could be correlated to the very large differences of 10–15 fold, which are seen in the incidence of clinical prostate cancer in countries with a Western and Eastern lifestyle.

IS FOCAL PROSTATE CANCER STIMULATED BY ANDROGENS?

An answer to this question is crucial for the understanding of the possible promotional role of androgens in this disease. Unfortunately, no clear cut evidence exists to confirm or to refute this working hypothesis. At the level of the genome promotion is likely to be a multistep event. Mutations at the level of the androgen receptor are likely to be involved. Some of these mutations could lead to androgen hypersensitivity as postulated by Visakorpi and co-workers[16,17]. Androgen sensitivity, which occurs through androgen receptor amplification described by these authors, could indeed be a molecular event which may be shown to play a role in the progression of focal disease. Anecdotal evidence exists concerning the development of very aggressive prostate cancer in young body builders and athletes using androgenic hormones. Such cases are described by Roberts and Essenhigh[18] and Wemyss-Holden and colleagues[19]. Exogenous androgen supplementation in castrated or hypopituitaric men can induce and maintain benign prostatic hyperplasia. Statistical information on these issues is unavailable, the evidence remains anecdotal but is strong in the reported cases.

Wallace and co-workers[20] used 200 mg of testosterone enanthate intramuscularly weekly as a potential contraception regimen and produced significant increases in plasma levels of androgens. Interestingly, these authors did not find an elevation of the androgen-dependent protein prostate-specific antigen which is a very sensitive marker of the presence of prostate cancer. This observation, together with the knowledge that prostate-specific antigen rises constantly and at rates that are reasonably well understood in cases of prostate cancer, may allow the use of prostate-specific antigen as a monitor in situations where androgen treatment or substitution is warranted.

CONCLUSION

There is no doubt that established prostate cancer is stimulated by androgens. The crucial question whether precursor lesions are stimulated to progress to the clinical disease cannot be answered at this time. The possibility can, however, not be excluded that small differences in circulating androgen levels may explain large geographic variations of prostate cancer. The possibility that promotion of precursor lesions is stimulated by androgens cannot be excluded. Also, a precise definition of what clinical carcinoma really is, is missing. The possibility that the 4–5% of screen-detectable prostate cancers in the age group between 55 and 75 could be stimulated to grow at an even earlier stage of their development is alarming.

At this moment it must be concluded that the role of endogenous and exogenous androgens, with respect to the carcinogenesis of prostate cancer and especially the promotion of small lesions to clinically relevant disease, is at least permissive. A stimulatory effect cannot be excluded, especially since explanatory mechanisms are available. Therefore, androgen substitution, if indicated, should not lead to superphysiological plasma levels of androgenic steroids.

REFERENCES

1. Manni, A., Santen, R. J., Boucher, A. E., Lipton, A., Harvey, H., Simmonds, M., Gordon, R., Rohner, T., Drago, J., Wettlaufer, J. and Glode, L. M. (1987). Androgen depletion and repletion as a means of potentiating the effects of cytotoxic chemotherapy in advanced prostate cancer. *J. Steroid. Biochem.*, **27** (1–3), 551–6

2. Fowler, J. E. and Whitmore, W. F. (1981). The response of metastatic adenocarcinoma of the prostate to exogenous testosterone. *J. Urol.*, **126**, 372–5

3. Goldenberg, S. L., Bruchovsky, N., Gleave, M. E., Sullivan, L. D. and Akakura, K. (1995). Intermittent androgen suppression in the treatment of prostate cancer: a preliminary report. *Urology*, **45** (5), 839–45

4. Weerden, W. M. van, Steenbrugge, G. J. van, Kreuningen, A. van, Moerings, E. P. C. M., Jong, F. H. de and Schröder, F. H. (1991). Assessment of the

critical level of androgen for growth response of transplantable human prostatic carcinoma (PC-82) in nude mice. *J. Urol.*, **145**, 631–4

5. Weerden, W. M. van, Steenbrugge, G. J. van, Jong, F. H. de and Schröder, F. H. (1989). Relevance of low androgen levels and adrenal androgens in the growth of transplantable human prostatic carcinomas. *Urol. Res.*, **17**, (5), 342 (Abstract 64)

6. Weerden, W. M. van, Kreuningen, A. van, Elissen, N. M. J., Jong, F. H. de, Streenbrugge, G. J. van and Schröder, F. H. (1992). Effects of adrenal androgens on the transplantable human prostate tumor PC-82. *Endocrinology*, **131**(6), 2909–13

7. Sakr, W. A., Grignon, D. J., Crissman, J. D., Heilbrun, L. K., Cassin, B. J., Pontes, J. J. and Haas, G. P. (1994). High grade prostatic intraepithelial neoplasia (HGPIN) and prostatic adenocarcinoma between the ages of 20–69: an autopsy study of 249 cases. *In Vivo*, **8**(3), 439–43

8. Franks, L. M. (1954). Latent carcinoma of the prostate. *J. Path. Bact.*, **68**, 603–16

9. Breslow, N., Chan, C. W., Dhom, G., Drury, R. A. B., Franks, L. M., Gellei, B., Lee, Y. S., Lundberg, S., Sparke, B., Sternby, N. H. and Tulinius, H. (1977). Latent carcinoma of prostate at autopsy in seven areas. *Int. J. Cancer*, **20**, 680–8

10. Schröder, F. H., Damhuis, R. A. M., Kirkels, W. J., Koning, H. J. de, Kranse, R., Nijs, H. G. T. and Blijenberg, B. G. (1996). European randomized study of screening for prostate cancer – the Rotterdam pilot studies. *Int. J. Cancer*, **65**, 145–51

11. Akazaki, K. and Stemmermann, G. N. (1973). Comparative study of latent carcinoma of the prostate among Japanese in Japan and Hawaii. *J. Nat. Cancer Inst.*, **50**, 1137–44

12. Adlercreutz, H. (1990). Western diet and Western diseases: some hormonal and biochemical mechanisms and associations. *Scand. J. Clin. Lab. Invest.*, **50** (Suppl. 201), 3–23

13. Hämäläinen, E., Adlercreutz, H., Puska, P. and Pietinen, P. (1984). Diet and serum sex hormones in healthy men. *J. Steroid. Biochem.*, **20**, 459–64

14. Jong, F. H. de, Oishi, K., Hayes, R. B., Bogdanovicz, J., Raatgever, J., Maas, P. J. van der, Yoshima, O. and Schröder, F. H. (1991). Peripheral hormone levels in controls and patients with prostatic cancer or benign prostatic hyperplasia: results from the Dutch–Japanese case–control study. *Cancer Res.*, **51**(13), 3445–50

15. Carter, H. B., Pearson, J. D., Metter, E. J., Chan, D. W., Andres, R., Fozard, J. L., Rosner, W. and Walsh, P. C. (1995). Longitudinal evaluation of serum androgen levels in men with and without prostate cancer. *Prostate*, **27**, 25–31

16. Visakorpi, T., Hyytinen, E., Koivisto, P., Tanner, M., Keinänen, R., Palmberg, Chr., Palotie, A., Tammela, T., Isola, J. and Kallioniemi, O.-P. (1996). *In vivo* amplification of the androgen receptor gene and progression of human prostate cancer. *Nat. Genet.*, **9**(4), 401–406

17. Visakorpi, T., Kallioniemi, A. H., Svyänen, A.-C., Hyytinen, E. R., Karhu, R., Tammela, T., Isola, J. J. and Kallioniemi, O.-P. (1995). Genetic changes in primary and recurrent prostate cancer by comparative genomic hybridization. *Cancer Res.*, **55**, 342–7

18. Roberts, J. T. and Essenhigh, D. M. (1986). Adenocarcinoma of prostate in 40-year-old body-builder. *Lancet*, **8509** (Vol II), 742

19. Wemyss-Holden, S. A., Hamdy, F. C. and Hastie, K. J. (1994). Steroid abuse in athletes, prostatic enlargement and bladder outflow obstruction – is there a relationship? *Br. J. Urol.*, **74**, 476–8

20. Wallace, E. M., Pye, S. D., Wild, S. R. and Wu, F. C. W. (1993). Prostate-specific antigen and prostate gland size in men receiving exogenous testosterone for male contraception. *Int. J. Androl.*, **16**, 35–40

DISCUSSION

Bain: Are you aware of any data indicating that men who have been treated with testosterone do indeed have an increased incidence of prostate cancer over the years or a hastening of the progress of prostate cancer? What data do we actually have relating to androgen treatment and prostate cancer in a clinical setting?

Schröder: I think that of all the data we have, the follow-up is very short and the numbers are very small. Therefore, it is at this moment impossible to deduce from prospective studies whether there is any effect on prostatic carcinogenesis. Considering the statistics of this process, we cannot expect to see any conclusive effects, unless we have very large numbers of substituted males followed-up in a careful way.

Tenover: I would like to draw attention to an issue we find difficult in our supplementation research. We detected two prostate cancers in a prospective study that comprised a total of 70 men. One cancer was picked up because of the frequent rectal exams performed as part of the study. The other case was a man who had recurrent problems with fever, since he had had the same problem 20 years ago, and at that time was diagnosed as having chronic prostatitis, he wanted us to biopsy him again. The biopsy revealed a small focal lesion. The question remains whether these cases had something to do with the testosterone treatment these men were receiving or whether they had pre-existed. The US prevalence for prostate cancer is about 12%, which means you can expect a number of men in a replacement study who probably have at least focal disease at the time of onset of androgen therapy. The issue is: how should we screen participants prior to inclusion in an androgen therapy trial? Should they have PSAs less than 4, normal ultrasounds, and normal digital rectals? Are biopsies then still necessary?

Orwoll: How comfortably can we rely on PSA monitoring during androgen treatment as a tool to assess the androgen action on neoplastic cells in the prostate?

Schröder: The answer might be that PSA is a rather solid marker for prostate cancer. However, a normal PSA (and normal is below 4) does not in practice exclude it. In the largest series of radical prostatectomy, that of Walsh at Johns Hopkins[1] 40% of cases had a PSA below 4. In our European screening study, we have seen that about 30% of the 170 prostate cancers detected had a PSA below 4. So using only PSA with a cut-off of 4 is not useful to exclude cancer. One has to also do a rectal examination, which will correct for much of the error of the PSA. Furthermore, we might think of lowering the PSA cut-off to 2, and actually we have proposed this to the Dutch Cancer Society, to streamline the recommendations for screening. We must not forget that the PSA is the most sensitive parameter as compared with rectal examination and ultrasonography. We will miss about 8% of cancers with a PSA below 2. In the clinical trial setting we must only include men with a normal PSA in this sense, and we might keep rectal examination and ultrasonography in reserve.

Bergink: As a matter of speculation, in order to prevent development or promotion of prostate cancer, could it be worthwhile to add progesterone or progestogens to androgen treatment? This could turn off endogenous androgen production while at the same time the supplementation maintains the physiological testosterone levels?

Schröder: This seems to be a very fascinating idea.

de Wied: Is there a relationship between the size of the prostate and the circulating levels of androgens?

Thijssen: Looking at the size of the prostate, you will discover that in a Western population it increases with age. In non-Western populations, with very similar testosterone levels in the plasma, you will, however, find smaller prostates. There is no clear correlation and prostate size is determined by more than just the sex steroids.

Orwoll: Coming back to PSA as a tool to evaluate prostate changes under treatment: suppose that we observe that a man who had a PSA of 3 at baseline gets a PSA of 4.5 after administration of androgens. However, there is no palpable abnormality in the prostate. What would you suggest?

Schröder: I would suggest doing a biopsy. Of course, the man should be informed of the risks of doing a biopsy (if a tumor is found the treatment may mean onset of impotence or urinary incontinence), and the risks of not doing a biopsy. But personally, I would opt for a biopsy.

Orwoll: This is really a terrible problem, because chances are not small that you find an abnormality that may never be clinically significant in that man. For research purposes, the tumor is going to be clearly counted as an adverse effect of the androgen replacement therapy, whereas you do not know if that is actually the case.

Morales: Doing biopsies or not is a very controversial issue. We published a paper in the *Journal of Urology* on the use of supplemental testosterone in patients with impotence in which we performed no biopsies prior to treatment[2]. A group from Boston later wrote that it was mandatory to do biopsies in everybody who was going to have supplemental androgen therapy[3]. In the perspective of the problems outlined during this discussion, we do not agree with that. Nevertheless, the situation illustrates the controversy that is currently existing on the topic.

REFERENCES

1. Partin, A. W., Yoo, J., Carter, H. B., Pearson, J. D., Chan, D. W., Epstein, J. I., Walsh, P. C. (1993). The use of prostate specific antigen, clinical stage and Gleason score to predict pathological stage in men with localized prostate cancer. *J Urol.,* **150**, 110–14

2. Morales, A., Johnston, B., Heaton, J. W. P., Clark, A. (1994). Oral androgens in the treatment of hypogonadal impotent men. *J. Urol.,* **152**, 1115–18

3. Bruning, C. O., DeWolf, W. C., Morgentaler, A. (1995). Occult prostate cancer in hypogonadal men prior to testosterone therapy. *J. Urol.,* **153**, 330A

14

Androgen supplementation in practice: the treatment of erectile dysfunction associated with hypotestosteronemia

A. Morales

A variety of endocrinopathies have been implicated in the development of male erectile failure. In some, such as diabetes mellitus, the development of microvascular disease and autonomic and peripheral neuropathy lead to failure of the erectile mechanisms. In others, a causal relationship is purely circumstantial and remains unproven. However, there are two recognized alterations of the hypothalamus–pituitary–gonadal axis clearly associated with the development of erectile dysfunction: hypogonadism (hypotestosteronemia) and hyperprolactinemia. This section will deal exclusively with the therapy of hypogonadal impotence. Despite the infrequency (see below) of hypogonadal erectile dysfunction, there is a pervasive and indiscriminate use of exogenous androgens to men complaining of erectile dysfunction. This is due, in large part, to the availability of medication capable of correcting the biochemical parameters in the serum of impotent men. Currently available drugs fulfil fairly well the ideal for a pharmacological agent for the treatment of some hormonal abnormalities associated with impotence. The criteria commonly accepted include:

(1) Effective,

(2) Useful 'on demand',

(3) Free of side-effects,

(4) Easy to administer (oral or topical), and

(5) Affordable.

The estimates for either primary or secondary hypogonadism as a cause of impotence vary but are in the vicinity of 10%[1]. Based upon a collective experience it is commonly believed that hypogonadal impotence should exhibit both insufficient erectile quality for intromission, associated with a decrease in libido, as well as clear evidence of abnormally low levels of serum total testosterone and free testosterone, while hyperprolactinemia may or may not be present and the levels of luteinizing hormone (LH) may be normal or elevated. Other features often added to these criteria are testicular size, absence of vasculopathy and psychopathology. However, recent observations by us and others[2] with regard to the futility of serum testosterone and other endocrine markers in the diagnosis and monitoring of erectile dysfunction, even in biochemically hypogonadal men, are forcing us to question this system of patient assessment. Nevertheless, the standard criteria should be maintained until better markers of hypogonadism become available. It is clear that a fundamental problem with hypogonadism is a matter of definition since there is no question that agonadal (usually for surgical reasons) men benefit greatly from androgen supplementation. The borderline biochemical abnormalities are the hardest to interpret and the results of treatment the most difficult to evaluate.

The majority of patients complaining of impotence who are found, after investigation, to be hypogonadal are men over the age of 55 years. It is well established that there is a progressive androgen decline in the aging man (PADAM) which has been erroneously equated to the menopause in women. Hypoestrogenism leading to menopause is a well recognized clinical situation while PADAM and its consequences remain a controversial issue. It is known, however, that in men advancing age is associated with a decrease in both total and free plasma testosterone, an increase in both the plasma levels of sex hormone-binding globulin and a sensitivity of gonadotropins to androgen feedback as well as a flattening of the circadian variations in plasma testosterone[3]. It remains to be defined whether the same biochemical criteria (plasma hormone levels) can be applied to the older as to the young male population. To complicate matters it is not known if target organ sensitivities to androgens are the same for old as for young men. The implications for therapy, of these poorly defined issues, are obvious.

Although it is evident that testosterone supplementation for the hypogonadal or agonadal young man is usually mandatory, the situation may be different for the older man. The adverse effects of hypotestosteronemia include osteoporosis, decreases in lean body mass, lower red cell counts and hemoglobin and alterations in the sexual response, mood and cognitive function. There is increasing evidence that some of these abnormalities may be reversed with testosterone supplementation[4].

Currently, there are no published practical guidelines for treatment of hypogonadal men with erectile dysfunction but there are many contradictory opinions on the effects of androgen administration to these men. The National Institutes of Health Conference on Impotence[5] proposed that 'for some patients with hypogonadism androgen replacement therapy may sometimes be effective in improving erectile function'.

There is no evidence that supplemental testosterone is of any benefit to eugonadal men. On the contrary, the evidence suggests that men with normal or only marginally low levels of serum testosterone are poor responders to exogenous androgenic steroids. Profound hypogonadism responds best. The practical pros and cons of commonly used preparations are listed in Table 1.

A great deal of concern has been expressed about their effect in a number of metabolic and organ-system functions. They include: body composition; bone metabolism; lipid profile; hematological and bio-chemical parameters; liver function (cholestatic hepatitis and hepatic cancer); and prostate (promotion of benign prostate hyperplasia and enhancement of carcinoma).

Physicians are particularly aware of these concerns because the various forms of testosterone are prescribed to a population of men usually over 50 years of age when alterations of lipid metabolism and effects on the prostate gland may carry serious implications. Early studies have shown that testosterone supplementation in aging males results in an increase in erythropoiesis and lean body mass. More recent studies by Tenover[6] have confirmed these findings but have further shown that patients receiving exogenous testosterone exhibit a significant decline in total cholesterol and low-density lipoprotein cholesterol. Her observations have been further supported by Friedl and colleagues[7], Marin and colleagues[8] and Barret-Connor[9]. In addition the recent study of Phillips and co-workers[10], investigating parameters of sex hormones and their influence in the occurrence of myocardial infarction, showed that 'the correlations found in this study between testosterone and the degree of coronary artery disease and between testosterone and other risk factors for myocardial infarction raise the possibility that in men hypotestosteronemia may be a risk factor for coronary arteriosclerosis'. Therefore, the currently available evidence does not support concerns about negative effects on lipid profile and other

Table 1 Available androgen preparations with their recommended doses and most prominent advantages and disadvantages

Administration form generic name / *Commercial name*	*Recommended dose*	*Advantages*	*Disadvantages*
Injectable esters			
testosterone cypionate		well-tolerated	i.m. injection
Depot–Testosterone®	200 mg i.m. every 2 weeks	effective	'roller-coaster' effect
PMS–Testosterone Enanthate®	200 mg i.m. every 2 weeks	universally recommended	tachyphylaxis
testosterone enanthate			
Delatestryl®	200 mg i.m. every 2 weeks		
Alkylated oral preparations			
fluoxy-materone		oral	limited effectiveness
Halotestin®	10 mg t.i.d.		serious liver toxicity
methyl-testosterone			
Metandren®	10 mg t.i.d.		
Virilon®	10 mg t.i.d.		
Non-alkylated oral preparations			
testosterone undecanoate		equivalent to i.m. esters	t.i.d. administration
Andriol®	120–160 mg daily in three divided doses after meals	no 'roller coaster' effect oral	costly
Transdermal			
testosterone transdermal system		easy to use	daily application
Androderm®	5 mg daily (two patches)	circadian curve	skin irritation
Transscrotal			
testosterone patch		easy to use	scrotal shaving
Testoderm®	1 mg daily (one patch)	circadian rhythm	daily application

i.m., intramuscular; t.i.d., three times per day

risk factors of cardiovascular disease for patients receiving exogenous testosterone supplementation.

Urologists most frequently express worries about the potential effect of androgen stimulation on the prostate. By now, there is a serious body of information on which to base an opinion in this regard. Experimental evidence has been presented showing that canine prostatic epithelial cells cultured *in vitro* in the presence or absence of several sex steroids exhibit the same growth pattern regardless of whether steroids are present or not. The same studies demonstrated that the proliferative response was dependent on the time and concentration of serum. Therefore, these investigators[11] concluded that humoral factors, other than steroids may be of importance in the activation of epithelial cells leading to the development of prostatic hyperplasia. Short term clinical studies in aging men treated with injectable testosterone enanthate (100 mg every 3 weeks for 3 months) did not induce a significant increase in prostate size or the amount of post-void residual but there was a significant increase in the prostate-specific antigen which, however, remained within normal limits. The oral undecanoate form of testosterone has been evaluated in long-term clinical trials. In a controlled study it was found to result in a modest (12%) but statistically significant increase in the gland volume but no changes in the values of serum prostate-specific antigen between the pre-, during and post-treatment readings[12]. These and other studies are reassuring but not conclusive as to the safety of supplemental testosterone on the biological behavior of the gland. Despite the well-recognized appearance of PADAM and the early, reassuring safety results with supplemental androgens, their use in this situation remains controversial. Controlled studies with larger cohorts and long periods of treatment and follow-up are mandatory to elucidate these important issues[13].

Anyone treating men with erectile dysfunction, should be familiar with the individual merits and drawbacks of the currently available testosterone preparations. The most acceptable to the patients are the oral compounds. Of these, methyl-testosterone requires repeated daily administration, results in supraphysiological levels of circulating total testosterone with dissociated values of free testosterone[14]. Its major drawbacks include hepatic toxicity and erratic absorption[15]. The other oral compound is testosterone unde-canoate, which also requires daily doses at 8 h intervals and results in more physiological levels of both serum total and free testosterone. Its lymphatic absorption requires it to be ingested with meals; this circumvents the first hepatic passage thus preventing liver toxicity and rapid aromatization. In our experience it produces better subjective responses than its methyl counterpart. Both oral forms offer the advantage of stable levels of serum testosterone, thus preventing the ups and downs frequently observed with the injectable esters. Testosterone patches are relatively new and represent an appealing concept. In a limited study of four patients, McClure and co-workers[16] reported that their use achieved normal levels of both testosterone and dihydrotestosterone, resulted in no hepatic toxicity and there were no appreciable changes in the lipid profile although the high-density lipoproteins decreased slightly; this was a short study of 12 weeks. Longer studies using the transdermal administration route confirmed the safety of these preparations[17]. The application of a patch results in a physiological curve of serum testosterone that imitates the circadian serum levels of the hormone. Drawbacks of transdermal testosterone include the need for daily application, for some formulations, to the scrotal skin where the absorption of the drug is particularly effective. There is also the need to shave the area frequently in order to facilitate its transdermal delivery. More recent preparations do not require scrotal application.

Testosterone injectable esters were recommended as the only acceptable form by the panel of the Consensus Conference on Impotence[5]. This view, of course, was based on concerns about the toxicity of certain oral preparations and disregard for new forms of delivery. The injectable drugs (cypionate, enanthate or propionate) are generally well tolerated and remain the gold standard for effectiveness. Drawbacks of these preparations include the need for periodic intramuscular administration (every 2–3 weeks) and the high activity observed in the first 10 days after administration, which is followed by a noticeable decrease towards the end of the cycle[18]. Patients find this 'roller-coaster' effect disturbing and unpleasant.

Finally, the cost of the drugs is a factor to be taken into consideration. The figures provided here are approximated and in Canadian dollars. On a comparative basis, oral androgen supplementation with either testosterone undecanoate (120 mg/day, $ 3.40/day) or methyl testosterone (30 mg/day, $ 3.68/day) or transdermal systems ($ 4.00/day; currently not yet available in Canada) are considerably more expensive than the intramuscular androgen preparations (300 mg every 3 weeks for cypionate: $ 0.31/day; enanthate $ 0.36/day and propionate $ 0.20/day). It must be considered, however, that if the cost of the clinic visit is included (one injection per visit

every 3 weeks, approximately $ 40.00 × 3 visits over 60 days = $ 2.00/day)
the cost difference between the various formulations is narrower.

We would like to conclude by making reference to the guidelines
for androgen supplementation approved by the Canadian Urological
Association, which accord with our clinical experience. These guidelines
are easy to follow and practical. The essence of the recommendations is as
follows:

(1) Patients receiving supplemental testosterone therapy should have a
clear indication for this therapy based on history, physical examination
and laboratory assessment. The diagnosis of hypotestosteronemia is
made by finding levels of serum total or, preferably free, testosterone
below the lower limits of normal. This should be demonstrated on two
or more occasions from morning blood samples. It is possible, though
not yet confirmed, that a hypoandrogenic state exists when the testosterone level is low–normal in the presence of an elevated serum LH.

(2) No patient is too old to start testosterone therapy if it is clearly indicated.
Hypogonadal patients usually require testosterone therapy for life.

(3) Patients with suspected secondary hypogonadism (of hypothalamic–
pituitary origin) should not receive testosterone treatment until their
endocrine investigation has been completed.

(4) Prior to the initiation of testosterone therapy, all patients should have
a digital rectal examination performed and a serum prostate-specific
antigen level measured.

(5) Mild benign prostatic hypertrophy is a relative, not an absolute
contraindication to testosterone therapy. Hypogonadal patients with
few or no urinary tract obstructive symptoms may be suitable for
testosterone therapy, but patients with advanced obstructive symptoms
are not.

(6) Testosterone treatment may result in increased estrogen levels and is
therefore contraindicated in men with breast cancer.

(7) Known prostatic cancer is a contraindication for testosterone treatment. Patients suspected of having prostatic cancer by either digital
rectal examination or prostate-specific antigen should be fully investigated.

(8) Oil-based testosterone injectables and oral testosterone undecanoate are the recommended agents to use because of their safety. The 17-alkylated steroids are not recommended because of their potential to cause hepatoxicity.

(9) Initially, patients should be followed on a 6-monthly basis. At these visits, the following should be undertaken:

(a) history of response to therapy or adverse effects particularly with regard to urinary obstructive symptoms;

(b) digital rectal examination if older than 40 years of age;

(c) prostate-specific antigen if older than 40 years of age.

Patients who remain stable may subsequently be followed annually at which time lipid profile, hemoglobin and serum calcium could also be considered.

(10) Serum levels of testosterone in treated patients will fluctuate considerably, particularly if testosterone is given by intramuscular injections. Regardless of serum levels, clinical response is a better guide to the dose required.

REFERENCES

1. Spark, R. F., White, R. A. and Connolly, P. B. (1980). Impotence is not always psychogenic: newer insights into hypothalamic–pituitary–gonadal dysfunction. *J. Am. Med. Assoc.*, **243**, 750–5

2. Johnson, A. R. III and Jarow, J. P. (1992). Is routine endocrine testing for impotent men necessary? *J. Urol.*, **147**, 1542–4

3. Vermeulen, A. (1991). Clinical review 24: androgens in the aging male. *J. Clin. Endocrinol. Metab.*, **73**, 221–6

4. Tenover, J. S. (1994). Androgen administration to aging men. *Endocrinol. Metab. Clin. North Am.*, **23**, 877–92

5. NIH Consensus Conference (1983). Impotence. NIH Consensus Development Panel on Impotence. *J. Am. Med. Assoc.*, **270**, 83–7

6. Tenover, J. S. (1992). Effects of testosterone supplementation in the aging male. *J. Clin. Endocrinol. Metab.*, **75**, 1092–8

7. Friedl, K. E., Hannan, C. L., Jones, R. E. and Playmate, S. R. (1990). High-density lipoprotein cholesterol is not decreased if an aromatizable androgen is administered. *Metabolism,* **39**, 69

8. Marin, P., Holmang, S., Jonsson, L., Sjostrom, L., Kvist, H., Holm, G., Goran, L. and Björntorp, P. (1992). The effects of testosterone treatment on body composition and metabolism in middle-aged obese men. *Int. J. Obesity,* **16**, 991

9. Barrett-Connor, E. (1992). Lower endogenous androgen levels and dyslipidemia in men with non-insulin-dependent diabetes mellitus. *Ann. Intern. Med.,* **117**, 807

10. Philips, G. B., Pinkernell, B. H. and Jing, T. (1994). The association of hypotestosteronemia with coronary artery disease in men. *Arterioscler. Thromb.,* **14**, 701

11. Chevalier, S., Bleau, G., Roberts, K. D. and Chapdelaine, A. (1984). Non-steroidal serum factors involved in the regulation of the proliferation of canine prostatic epithelial cells in culture. *Prostate,* **5**, 503

12. Behre, H. M., Bohmeyer, J. and Nieschlag, E. (1994). Prostate volume in testosterone treated and untreated hypogonadal men in comparison to age-matched controls. *Clin. Endocrinol.,* **40**, 341–6

13. Vermeulen, A. and Kaufman, J. M. (1995). Ageing of the hypothalamic–pituitary–testicular axis in man. *Horm. Res.,* **43**, 25–8

14. Morales, A., Johnston, B., Heaton, J. P. W. and Clark, A. (1994). Oral androgens in the treatment of hypogonadal impotent men. *J. Urol.,* **152**, 115–8

15. Wilson, J. D. and Griffith, J. D. (1980). The use and misuse of androgens. *Metabolism,* **29**, 1278

16. McClure, R. D., Oses, R. and Ernest, M. L. (1991). Hypogonadal impotence treated by transdermal testosterone. *Urology,* **37**, 224

17. Orwoll, E., Oviatt, S. and Biddle, J. (1992). Transdermal testosterone supplementation in normal older men. Abstracts of the 74th Annual Meeting of the Endocrine Society, p. 319

18. Snyder, P. J. and Lawrence, D. A. (1980). Treatment of male hypogonadism with testosterone enanthate. *J. Clin. Endocrinol. Metab.,* **51**, 1335–41

DISCUSSION

Orwoll: I am interested in your recommendations for follow-up. By suggesting that rectal examinations should be done every 6 months after beginning the therapy, it would seem as if you are implying that prostatic disease might be more common in people supplemented with testosterone for hypogonadism than in the general population of eugonadal men.

Morales: My feeling is that these hypogonadal individuals will probably have a prostate that is smaller, and by giving the replacement therapy, it is going to achieve the anticipated size for the respective age and level of androgens. However, since the literature is confusing, we felt that it was probably appropriate to err on the safe side: do it every 6 months for the first 2 years and see whether the prostate is really going to grow. We are now starting to do uroflows and to evaluate if indeed there is development of obstructive symptoms. So far, we do not have sufficient information on these issues, and therefore we err on the safe side.

Orwoll: It seems as if your recommendations vary on the basis of age. Would you not do digital rectal examinations at 6 month intervals after beginning therapy for those men less than 40 years old?

Morales: Personally, I do not think it is necessary. The recommendations from both the American and the Canadian Urological Associations in regard to prostatic cancer indicate that you should start the screening procedures after the age of 50, and only under special circumstances after the age of 40 (family history of carcinoma, vasectomy). I do not see any justification of putting a 25-year-old agonadic man through all this rigmarole of tests.

Vermeulen: Do you agree with this, Dr Schröder?

Schröder: Yes, I do.

Tenover: Data were presented at the International Andrology Workshop last February (1995) which showed that in older hypogonadal men, aged over 60, PSA levels may not be as predictive as perhaps in the general population. Biopsies had been performed on a large group of older hypogonadal men before they were given therapy, and the investigation found a very high rate of prostate cancer *in situ*.

Schröder: The issue of prostate cancer in hypogonadal men is a very interesting one. There are descriptions that if prostate cancer occurs in a man with low testosterone levels, the cancer may be very aggressive and lead to the death of the patient very soon. The argumentation for this has been that in fact the tumor had already been 'treated' by the existence of the low androgen levels. This is something to be conscious of, when you detect prostatic cancer in a hypogonadal man.

Morales: Could I pose a more general question to the audience? Until recently, it was very common to measure total testosterone, at least in urological areas. I am beginning to question the value of total serum testosterone. I would think it is more important to have the values of bioavailable testosterone.

Vermeulen: Well of course, bioavailable testosterone is a better parameter, but if the patient is not obese and no obvious interfering factors enter the picture, total testosterone is a very reliable parameter.

Bain: At the Mount Sinai hospital in Toronto, we had been using total testosterone for quite some years until we introduced free testosterone assessments. We tested quite a number of patients for both total and free testosterone. Eventually, we were much more satisfied with the free testosterone, because the total testosterone did not always fit the clinical situation, in contrast to the free testosterone. We also detected situations where total testosterone was on the lower side, but free testosterone consistently normal. Therefore, we do not determine total testosterone any more, but only free testosterone.

Haffner: We measure routinely both total and free testosterone, partially because the journals request it. There are now some good free testosterone assays which are actually very precise. But until recently, the imprecision of the free testosterone was such, that you were better off measuring SHBG and total T, and then with the help of computer programs figured out the estimated free testosterone. I think in most clinical work, total testosterone is still probably adequate.

Longcope: We have heard in this workshop about osteoporosis in hypogonadal men, and I did not hear Dr Morales mention anything about spine density measurements in hypogonadal and middle-aged men. I wondered whether discontinuation of the therapy because of lack of clinical improvement would not be a disservice to the men concerned because they may end up with osteoporosis?

Orwoll: The issue of bone density is a very important one. It is clear that hypogonadism is a risk factor for the development of osteoporosis, which is a subclinical disease until it is manifest by fractures. I would suggest that a measurement of bone density is a very important part of the decision whether to continue testosterone therapy even in the absence of the alleviation of clinical symptomatology. Apart from this issue, several of us were talking earlier about lipids and I wondered what you think of lipid concentrations being measured before therapy is begun. Can testosterone therapy worsen a pre-existing hyperlipidemia? And if you have a pre-existing hyperlipidemia, should it be treated, followed, or whatever, during therapy?

Morales: The information I heard during this workshop confirmed that there is not a major concern with lipids in relation to supplemental androgens. Perhaps it should be expanded to say that patients should have the baseline investigations, and then be followed over the following two years with determination of total cholesterol, HDL and LDL.

Schiavi: A number of the patients with hypoactive sexual desire, in terms of psychiatry, have adequate erectile function, even though the desire is low. How would you consider this category as far as intervention with androgen supplementation is concerned?

Morales: Presumably all these patients attend the clinic with complaints of erectile problems, caused by either a decrease in sexual desire or most frequently impotence. If they are found to be hypogonadal they are, in my view, candidates for supplementation. If they respond well, this was obviously a cause of their erectile difficulties. If they do not respond, it is very important to investigate co-morbidity.

Schiavi: The individuals who come to our human sexuality program complain of hypoactive sexual desire, but a number of them continue to have adequate erectile function.

Morales: My personal view would be to consider testosterone supplementation in these individuals, because there is more than just the erectile failure. Let me refer to the agonadal individuals: they have erections, but their sexual desire, their mood, their general behavior is very different, and they report this tremendous improvement when they receive supplementation. It is not just the fact that individuals are able to achieve vaginal penetration which is relevant, but the frequency, the quality, the orgasms, the production of the ejaculation, etc. count as well. All these factors are important to consider.

Oddens: Referring to Dr Tenover's earlier remarks that this intriguing field of androgen functioning has only been looked at in detail for over a decade now and taking into account that this decade of research has already yielded many indications, my last question of this workshop would be: do you think that in 10 years from now, androgen supplementation will be something common in the clinic?

Tenover: If you base your prediction on what has happened with estrogens in postmenopausal women, it will not be common practice. I mean, even though we have very clear insights into the risk/benefit ratio for postmenopausal use of estrogen in women, the use is not as widespread as one might expect. However, for the issue of androgen therapy, times may ultimately be different: the psyches of the population may be different and it may be different because we are dealing with men. I suspect that 10 years from now there will be a lot of intermittent androgen use, with men trying androgens for at least a short time period. But whether that means that men will stay on androgen therapy for long is a different question altogether.

Index